ELECTRICAL INSTALLATIONS UNDER NEC® 2002

Volume I: Residential

POCKET GUIDE TO ELECTRICAL INSTALLATIONS UNDER NEC® 2002

Volume I: Residential

CHARLES R. MILLER
Lighthouse Educational Services

National Fire Protection Association
Quincy, Massachusetts

Product Manager: Charles Durang
Editorial-Production Services: Omegatype Typography, Inc.
Interior Design: Omegatype Typography, Inc.
Composition: Omegatype Typography, Inc.
Cover Design: Natasha Bumbeck
Manufacturing Manager: Ellen Glisker
Printer: Rose Printing

NFPA No.: PGNECRES-02
ISBN: 0-87765-457-3
Library of Congress Control Number: 2001093272

Printed in the United States of America
05 04 03 02 01   5 4 3 2 1

IMPORTANT NOTICES AND DISCLAIMERS OF LIABILITY

This *Pocket Guide* is a compilation of extracts from the 2002 edition of the *National Electrical Code®* (*NEC®*). These extracts have been selected and arranged by the Author of the *Pocket Guide*, who has also supplied introductory material generally located at the beginning of each chapter.

The *NEC*, like all NFPA codes and standards, is a document that is developed through a consensus standards development process approved by the American National Standards Institute. This process brings together volunteers representing varied viewpoints and interests to achieve consensus on fire, electrical, and other safety issues. While the NFPA administers the process and establishes rules to promote fairness in the development of consensus, it does not independently test, evaluate, or verify the accuracy of any information or the soundness of any judgments contained in its codes and standards.

The NFPA and the Author disclaim liability for any personal injury, property or other damages of any nature whatsoever, whether special, indirect, consequential, or compensatory, directly or indirectly resulting from the publication, use of, or reliance on the *NEC* or this *Pocket Guide*. The NFPA and the Author also make no guaranty or warranty as to the accuracy or completeness of any information published herein.

In issuing and making the *NEC* and this *Pocket Guide* available, the NFPA and the Author are not undertaking to render professional or other services for or on behalf of any person or entity. Nor is the NFPA or the Author undertaking to perform any duty owed by any person or entity to someone else. Anyone using these materials should rely on his or her own independent judgment or, as appropriate, seek the advice of a competent professional in determining the exercise of reasonable care in any given circumstances.

For more important information and notices concerning the use of NFPA codes and standards, please consult the introductory materials accompanying all published versions of these documents.

NOTICE CONCERNING CODE INTERPRETATIONS

The introductory materials and the selection and arrangement of the *NEC* extracts contained in this *Pocket Guide* reflect the personal opinions and judgments of the Author and do not necessarily represent the official position of the NFPA (which can only be obtained through Formal Interpretations processed in accordance with NPFA rules).

®Registered trademarks of the National Fire Protection Association, Inc.

CONTENTS

CONTENTS

INTRODUCTION

The *Pocket Guide to Electrical Installations Under NEC®
2002, Volume I: Residential,* was compiled to provide a portable, handy reference source for field practitioners: installers, contractors, and inspectors. With its companion volume,
*Pocket Guide to Electrical Installations Under NEC® 2002,
Volume II: Commercial and Industrial,* it gathers together all
the *National Electrical Code®* rules you need for most standard installations, in the order you need them.

The code requirements are not organized in this
Pocket Guide in the same sequence found in the *NEC®*.
Rather, they are presented in an order similar to that of an
actual electrical system installation process. A few introductory chapters are followed by the requirements that pertain to temporary wiring. Next, the *Guide* covers certain
calculations that must be performed before the actual electrical installation (e.g., branch circuits, feeders, and services). The book then continues with provisions that are
essential for safeguarding persons and property from the
hazards arising from the use of electricity. Because this
book is literally a pocket guide, space does not permit an
in-depth reference to and explanation of *NEC* regulations.
Other instructional material, such as NFPA's *National Electrical Code Handbook,* can provide a better insight into and
understanding of the *Code.*

State or local agencies may—and often do—adopt
regulations above and beyond the *National Electrical
Code.* Local authorities can adopt the *NEC* exactly as written, or they can amend the *Code* or supplement it with
more stringent regulations. For example, additional restrictions may limit the number of lights or receptacles on a circuit. Obtain a copy of any additional requirements for your
city, county, and/or state.

These *Pocket Guides* are not intended to be a substitute
for the *NEC* itself, which should always be consulted to ensure full code compliance. The choice of material included in
this *Pocket Guide* was based on the problems and situations
most frequently encountered in the field, where it is not practical to carry bulky reference material. Remember, however,

that important provisions of the *Code,* such as the more specialized rules for special occupancies, special equipment, special conditions, or communications circuits, while not included, are important to know and understand, and should be consulted where appropriate.

It is hoped that this *Pocket Guide* will prove to be a valuable addition to your toolbox, glove compartment, briefcase, or, as the name implies, your pocket.

POCKET GUIDE TO ELECTRICAL INSTALLATIONS UNDER NEC® 2002

Volume I: Residential

CHAPTER 1

INTRODUCTION TO THE NATIONAL ELECTRICAL CODE

INTRODUCTION

This chapter contains provisions from Article 90—Introduction. Article 90 covers basic provisions essential for understanding and applying the *National Electrical Code®*. Besides providing the purpose of the *Code,* Article 90 specifies the document scope, describes how the *Code* is arranged, provides information on enforcement, and contains other provisions that are essential to the proper use and application of the *Code*. The first sentence states that "the purpose of this *Code* is the practical safeguarding of persons and property from hazards arising from the use of electricity." Safety, for both property and people, is the main objective for the provisions found in the *Code*. Compliance with these provisions and proper maintenance will result in an installation essentially free from electrical hazards but not necessarily efficient, convenient, or adequate for good service or future expansion of electrical use. Because of the limited space within this pocket guide, only part of Article 90 has been included. Reference the *2002 National Electrical Code* for all of the requirements.

　　Although state and local jurisdictions generally adopt the *Code* exactly as written, they may also amend the *Code* by incorporating local requirements. Therefore, obtain a copy of any additional rules and regulations for your area.

ARTICLE 90
Introduction

90.1 Purpose.

(A) Practical Safeguarding. The purpose of this *Code* is the practical safeguarding of persons and property from hazards arising from the use of electricity.

(B) Adequacy. This *Code* contains provisions that are considered necessary for safety. Compliance therewith and proper maintenance will result in an installation that is essentially free from hazard but not necessarily efficient, convenient, or adequate for good service or future expansion of electrical use.

90.4 Enforcement. This *Code* is intended to be suitable for mandatory application by governmental bodies that exercise legal jurisdiction over electrical installations, including signaling and communications systems, and for use by insurance inspectors. The authority having jurisdiction for enforcement of the *Code* has the responsibility for making interpretations of the rules, for deciding on the approval of equipment and materials, and for granting the special permission contemplated in a number of the rules.

By special permission, the authority having jurisdiction may waive specific requirements in this *Code* or permit alternative methods where it is assured that equivalent objectives can be achieved by establishing and maintaining effective safety.

This *Code* may require new products, constructions, or materials that may not yet be available at the time the *Code* is adopted. In such event, the authority having jurisdiction may permit the use of the products, constructions, or materials that comply with the most recent previous edition of this *Code* adopted by the jurisdiction.

90.5 Mandatory Rules, Permissive Rules, and Explanatory Material.

(A) Mandatory Rules. Mandatory rules of this *Code* are those that identify actions that are specifically required or prohibited and are characterized by the use of the terms *shall* or *shall not*.

(B) Permissive Rules. Permissive rules of this *Code* are those that identify actions that are allowed but not required, are normally used to describe options or alternative methods, and are characterized by the use of the terms *shall be permitted* or *shall not be required*.

(C) Explanatory Material. Explanatory material, such as references to other standards, references to related sections of this *Code,* or information related to a *Code* rule, is included in this *Code* in the form of fine print notes (FPNs). Fine print notes are informational only and are not enforceable as requirements of this *Code.*

90.8 Wiring Planning.

(A) Future Expansion and Convenience. Plans and specifications that provide ample space in raceways, spare raceways, and additional spaces allow for future increases in the use of electricity. Distribution centers located in readily accessible locations provide convenience and safety of operation.

(B) Number of Circuits in Enclosures. It is elsewhere provided in this *Code* that the number of wires and circuits confined in a single enclosure be varyingly restricted. Limiting the number of circuits in a single enclosure minimizes the effects from a short circuit or ground fault in one circuit.

CHAPTER 2

DEFINITIONS APPLICABLE TO DWELLING UNITS

INTRODUCTION

The definition of a word or phrase used in the *NEC* may differ from that found in a dictionary. In fact, some of these terms are not defined in a standard dictionary. Understanding the *Code* without a thorough understanding of the terms is difficult. Therefore, having an understanding of Article 100 is an important step toward proper application of the *Code*.

As stated in the scope of Article 100, only definitions essential to the proper application of the *Code* are included. Since commonly defined general terms can be found in most dictionaries, they are not included. Also not included are commonly defined technical terms. These terms can be found in a dictionary of electrical and electronic terms.

Since a number of terms in Article 100 do not apply to dwelling units, they are not included in this book. Likewise, because of this guide's compact design, other terms and fine print notes (FPN) are also omitted. Reference the *2002 National Electrical Code* for a complete listing.

Article 100 contains terms used in two or more articles. Other terms that pertain to one article may be defined in that particular article. For example, Article 680—Swimming Pools, Fountains, and Similar Installations includes terms that are not used anywhere else in the *Code*. Therefore, those terms are defined in Article 680 rather than in Article 100.

ARTICLE 100
Definitions

Scope. This article contains only those definitions essential to the proper application of this *Code*. It is not intended to include commonly defined general terms or commonly defined technical terms from related codes and standards. In general, only those terms that are used in two or more articles are defined in Article 100. Other definitions are included in the article in which they are used but may be referenced in Article 100.

Part I of this article contains definitions intended to apply wherever the terms are used throughout this *Code*. Part II contains definitions applicable only to the parts of articles specifically covering installations and equipment operating at over 600 volts, nominal.

I. GENERAL

Accessible (as applied to equipment). Admitting close approach; not guarded by locked doors, elevation, or other effective means.

Accessible (as applied to wiring methods). Capable of being removed or exposed without damaging the building structure or finish or not permanently closed in by the structure or finish of the building.

Accessible, Readily (Readily Accessible). Capable of being reached quickly for operation, renewal, or inspections without requiring those to whom ready access is requisite to climb over or remove obstacles or to resort to portable ladders, and so forth.

Ampacity. The current, in amperes, that a conductor can carry continuously under the conditions of use without exceeding its temperature rating.

Appliance. Utilization equipment, generally other than industrial, that is normally built in standardized sizes or types and is installed or connected as a unit to perform one or more functions such as clothes washing, air conditioning, food mixing, deep frying, and so forth.

Approved. Acceptable to the authority having jurisdiction.

Attachment Plug (Plug Cap) (Plug). A device that, by insertion in a receptacle, establishes a connection between the conductors of the attached flexible cord and the conductors connected permanently to the receptacle.

Authority Having Jurisdiction. The organization, office, or individual responsible for approving equipment, materials, an installation, or a procedure.

Automatic. Self-acting, operating by its own mechanism when actuated by some impersonal influence, as, for example, a change in current, pressure, temperature, or mechanical configuration.

Bathroom. An area including a basin with one or more of the following: a toilet, a tub, or a shower.

Bonding (Bonded). The permanent joining of metallic parts to form an electrically conductive path that ensures electrical continuity and the capacity to conduct safely any current likely to be imposed.

Bonding Jumper. A reliable conductor to ensure the required electrical conductivity between metal parts required to be electrically connected.

Bonding Jumper, Equipment. The connection between two or more portions of the equipment grounding conductor.

Bonding Jumper, Main. The connection between the grounded circuit conductor and the equipment grounding conductor at the service.

Branch Circuit. The circuit conductors between the final overcurrent device protecting the circuit and the outlet(s).

Branch Circuit, Appliance. A branch circuit that supplies energy to one or more outlets to which appliances are to be connected and that has no permanently connected luminaires (lighting fixtures) that are not a part of an appliance.

Branch Circuit, General-Purpose. A branch circuit that supplies two or more receptacles or outlets for lighting and appliances.

Branch Circuit, Individual. A branch circuit that supplies only one utilization equipment.

Branch Circuit, Multiwire. A branch circuit that consists of two or more ungrounded conductors that have a voltage between them, and a grounded conductor that has equal voltage between it and each ungrounded conductor of the circuit and that is connected to the neutral or grounded conductor of the system.

Building. A structure that stands alone or that is cut off from adjoining structures by fire walls with all openings therein protected by approved fire doors.

Cabinet. An enclosure that is designed for either surface mounting or flush mounting and is provided with a frame, mat, or trim in which a swinging door or doors are or can be hung.

Circuit Breaker. A device designed to open and close a circuit by nonautomatic means and to open the circuit automatically on a predetermined overcurrent without damage to itself when properly applied within its rating.

Concealed. Rendered inaccessible by the structure or finish of the building. Wires in concealed raceways are considered concealed, even though they may become accessible by withdrawing them.

Conductor, Bare. A conductor having no covering or electrical insulation whatsoever.

Conductor, Covered. A conductor encased within material of composition or thickness that is not recognized by this *Code* as electrical insulation.

Conductor, Insulated. A conductor encased within material of composition and thickness that is recognized by this *Code* as electrical insulation.

Conduit Body. A separate portion of a conduit or tubing system that provides access through a removable cover(s) to the interior of the system at a junction of two or more sections of the system or at a terminal point of the system.

Boxes such as FS and FD or larger cast or sheet metal boxes are not classified as conduit bodies.

Connector, Pressure (Solderless). A device that establishes a connection between two or more conductors or between one or more conductors and a terminal by means of mechanical pressure and without the use of solder.

Continuous Load. A load where the maximum current is expected to continue for 3 hours or more.

Controller. A device or group of devices that serves to govern, in some predetermined manner, the electric power delivered to the apparatus to which it is connected.

Cooking Unit, Counter-Mounted. A cooking appliance designed for mounting in or on a counter and consisting of one or more heating elements, internal wiring, and built-in or mountable controls.

Copper-Clad Aluminum Conductors. Conductors drawn from a copper-clad aluminum rod with the copper metallurgically bonded to an aluminum core. The copper forms a minimum of 10 percent of the cross-sectional area of a solid conductor or each strand of a stranded conductor.

Cutout Box. An enclosure designed for surface mounting that has swinging doors or covers secured directly to and telescoping with the walls of the box proper.

Dead Front. Without live parts exposed to a person on the operating side of the equipment.

Demand Factor. The ratio of the maximum demand of a system, or part of a system, to the total connected load of a system or the part of the system under consideration.

Device. A unit of an electrical system that is intended to carry but not utilize electric energy.

Disconnecting Means. A device, or group of devices, or other means by which the conductors of a circuit can be disconnected from their source of supply.

Dwelling Unit. One or more rooms for the use of one or more persons as a housekeeping unit with space for eating, living, and sleeping, and permanent provisions for cooking and sanitation.

Dwelling, One-Family. A building that consists solely of one dwelling unit.

Dwelling, Two-Family. A building that consists solely of two dwelling units.

Dwelling, Multifamily. A building that contains three or more dwelling units.

Enclosed. Surrounded by a case, housing, fence, or wall(s) that prevents persons from accidentally contacting energized parts.

Enclosure. The case or housing of apparatus, or the fence or walls surrounding an installation to prevent personnel from accidentally contacting energized parts or to protect the equipment from physical damage.

Energized. Electrically connected to a source of voltage.

Equipment. A general term including material, fittings, devices, appliances, luminaires (fixtures), apparatus, and the like used as a part of, or in connection with, an electrical installation.

Exposed (as applied to live parts). Capable of being inadvertently touched or approached nearer than a safe distance by a person. It is applied to parts that are not suitably guarded, isolated, or insulated.

Exposed (as applied to wiring methods). On or attached to the surface or behind panels designed to allow access.

Feeder. All circuit conductors between the service equipment, the source of a separately derived system, or other power supply source and the final branch-circuit overcurrent device.

Festoon Lighting. A string of outdoor lights that is suspended between two points.

Fitting. An accessory such as a locknut, bushing, or other part of a wiring system that is intended primarily to perform a mechanical rather than an electrical function.

Garage. A building or portion of a building in which one or more self-propelled vehicles can be kept for use, sale, storage, rental, repair, exhibition, or demonstration purposes.

Ground. A conducting connection, whether intentional or accidental, between an electrical circuit or equipment and the earth or to some conducting body that serves in place of the earth.

Grounded. Connected to earth or to some conducting body that serves in place of the earth.

Grounded, Effectively. Intentionally connected to earth through a ground connection or connections of sufficiently low impedance and having sufficient current-carrying capacity to prevent the buildup of voltages that may result in undue hazards to connected equipment or to persons.

Grounded Conductor. A system or circuit conductor that is intentionally grounded.

Grounding Conductor. A conductor used to connect equipment or the grounded circuit of a wiring system to a grounding electrode or electrodes.

Grounding Conductor, Equipment. The conductor used to connect the non–current-carrying metal parts of equipment, raceways, and other enclosures to the system grounded conductor, the grounding electrode conductor, or both, at the service equipment or at the source of a separately derived system.

Grounding Electrode Conductor. The conductor used to connect the grounding electrode(s) to the equipment grounding conductor, to the grounded conductor, or to both, at the service, at each building or structure where supplied from a common service, or at the source of a separately derived system.

Ground-Fault Circuit Interrupter. A device intended for the protection of personnel that functions to de-energize a circuit or portion thereof within an established period of time when a current to ground exceeds the values established for a Class A device.

Guarded. Covered, shielded, fenced, enclosed, or otherwise protected by means of suitable covers, casings, barriers, rails, screens, mats, or platforms to remove the likelihood of approach or contact by persons or objects to a point of danger.

Identified (as applied to equipment). Recognizable as suitable for the specific purpose, function, use, environment, application, and so forth, where described in a particular *Code* requirement.

In Sight From (Within Sight From, Within Sight). Where this *Code* specifies that one equipment shall be "in sight from," "within sight from," or "within sight," and so forth, of another equipment, the specified equipment is to be visible and not more than 15 m (50 ft) distant from the other.

Interrupting Rating. The highest current at rated voltage that a device is intended to interrupt under standard test conditions.

Labeled. Equipment or materials to which has been attached a label, symbol, or other identifying mark of an organization that is acceptable to the authority having jurisdiction and concerned with product evaluation, that maintains periodic inspection of production of labeled equipment or materials, and by whose labeling the manufacturer indicates compliance with appropriate standards or performance in a specified manner.

Lighting Outlet. An outlet intended for the direct connection of a lampholder, a luminaire (lighting fixture), or a pendant cord terminating in a lampholder.

Listed. Equipment, materials, or services included in a list published by an organization that is acceptable to the authority having jurisdiction and concerned with evaluation of products or services, that maintains periodic inspection of production of listed equipment or materials or periodic evaluation of services, and whose listing states that the equipment, material, or services either meets appropriate designated standards or has been tested and found suitable for a specified purpose.

Location, Damp. Locations protected from weather and not subject to saturation with water or other liquids but subject to moderate degrees of moisture. Examples of such locations include partially protected locations under canopies, marquees, roofed open porches, and like locations, and interior locations subject to moderate degrees of moisture, such as some basements, some barns, and some cold-storage warehouses.

Location, Dry. A location not normally subject to dampness or wetness. A location classified as dry may be temporarily subject to dampness or wetness, as in the case of a building under construction.

Location, Wet. Installations under ground or in concrete slabs or masonry in direct contact with the earth; in locations subject to saturation with water or other liquids, such as vehicle washing areas; and in unprotected locations exposed to weather.

Luminaire. A complete lighting unit consisting of a lamp or lamps together with the parts designed to distribute the light, to position and protect the lamps and ballast (where applicable), and to connect the lamps to the power supply.

Multioutlet Assembly. A type of surface, flush, or freestanding raceway designed to hold conductors and receptacles, assembled in the field or at the factory.

Outlet. A point on the wiring system at which current is taken to supply utilization equipment.

Overcurrent. Any current in excess of the rated current of equipment or the ampacity of a conductor. It may result from overload, short circuit, or ground fault.

Overload. Operation of equipment in excess of normal, full-load rating, or of a conductor in excess of rated ampacity that, when it persists for a sufficient length of time, would cause damage or dangerous overheating. A fault, such as a short circuit or ground fault, is not an overload.

Panelboard. A single panel or group of panel units designed for assembly in the form of a single panel, including buses and automatic overcurrent devices, and equipped with or without switches for the control of light, heat, or power circuits; designed to be placed in a cabinet or cutout box placed in or against a wall, partition, or other support; and accessible only from the front.

Plenum. A compartment or chamber to which one or more air ducts are connected and that forms part of the air distribution system.

Premises Wiring (System). That interior and exterior wiring, including power, lighting, control, and signal circuit wiring together with all their associated hardware, fittings, and wiring devices, both permanently and temporarily installed, that extends from the service point or source of power, such as a battery, a solar photovoltaic system, or a generator, transformer, or converter windings, to the outlet(s). Such wiring does not include wiring internal to appliances, luminaires (fixtures), motors, controllers, motor control centers, and similar equipment.

Qualified Person. One who has skills and knowledge related to the construction and operation of the electrical equipment and installations and has received safety training on the hazards involved.

Raceway. An enclosed channel of metal or nonmetallic materials designed expressly for holding wires, cables, or busbars, with additional functions as permitted in this *Code.* Raceways include, but are not limited to, rigid metal conduit, rigid nonmetallic conduit, intermediate metal conduit, liquidtight flexible conduit, flexible metallic tubing, flexible metal conduit, electrical nonmetallic tubing, electrical metallic tubing, underfloor raceways, cellular concrete floor raceways, cellular metal floor raceways, surface raceways, wireways, and busways.

Rainproof. Constructed, protected, or treated so as to prevent rain from interfering with the successful operation of the apparatus under specified test conditions.

Raintight. Constructed or protected so that exposure to a beating rain will not result in the entrance of water under specified test conditions.

Receptacle. A receptacle is a contact device installed at the outlet for the connection of an attachment plug. A single receptacle is a single contact device with no other contact device on the same yoke. A multiple receptacle is two or more contact devices on the same yoke.

Receptacle Outlet. An outlet where one or more receptacles are installed.

Remote-Control Circuit. Any electric circuit that controls any other circuit through a relay or an equivalent device.

Separately Derived System. A premises wiring system whose power is derived from a battery, from a solar photovoltaic system, or from a generator, transformer, or converter windings, and that has no direct electrical connection, including a solidly connected grounded circuit conductor, to supply conductors originating in another system.

Service. The conductors and equipment for delivering electric energy from the serving utility to the wiring system of the premises served.

Service Cable. Service conductors made up in the form of a cable.

Service Conductors. The conductors from the service point to the service disconnecting means.

Service Drop. The overhead service conductors from the last pole or other aerial support to and including the splices, if any, connecting to the service-entrance conductors at the building or other structure.

Service-Entrance Conductors, Overhead System. The service conductors between the terminals of the service equipment and a point usually outside the building, clear of building walls, where joined by tap or splice to the service drop.

Service-Entrance Conductors, Underground System. The service conductors between the terminals of the service equipment and the point of connection to the service lateral.

Service Equipment. The necessary equipment, usually consisting of a circuit breaker(s) or switch(es) and fuse(s) and their accessories, connected to the load end of service conductors to a building or other structure, or an otherwise designated area, and intended to constitute the main control and cutoff of the supply.

Service Lateral. The underground service conductors between the street main, including any risers at a pole or other structure or from transformers, and the first point of connection to the service-entrance conductors in a terminal box or meter or other enclosure, inside or outside the building wall. Where there is no terminal box, meter, or other enclosure, the point of connection is considered to be the point of entrance of the service conductors into the building.

Service Point. The point of connection between the facilities of the serving utility and the premises wiring.

Solar Photovoltaic System. The total components and subsystems that, in combination, convert solar energy into electrical energy suitable for connection to a utilization load.

Special Permission. The written consent of the authority having jurisdiction.

Structure. That which is built or constructed.

Switch, General-Use. A switch intended for use in general distribution and branch circuits. It is rated in amperes, and it is capable of interrupting its rated current at its rated voltage.

Switch, General-Use Snap. A form of general-use switch constructed so that it can be installed in device boxes or on box covers, or otherwise used in conjunction with wiring systems recognized by this *Code.*

Switch, Transfer. An automatic or nonautomatic device for transferring one or more load conductor connections from one power source to another.

Utilization Equipment. Equipment that utilizes electric energy for electronic, electromechanical, chemical, heating, lighting, or similar purposes.

Voltage (of a circuit). The greatest root-mean-square (rms) (effective) difference of potential between any two conductors of the circuit concerned.

Voltage, Nominal. A nominal value assigned to a circuit or system for the purpose of conveniently designating its voltage class (e.g., 120/240 volts, 480Y/277 volts, 600 volts).

The actual voltage at which a circuit operates can vary from the nominal within a range that permits satisfactory operation of equipment.

Voltage to Ground. For grounded circuits, the voltage between the given conductor and that point or conductor of the circuit that is grounded; for ungrounded circuits, the greatest voltage between the given conductor and any other conductor of the circuit.

Watertight. Constructed so that moisture will not enter the enclosure under specified test conditions.

Weatherproof. Constructed or protected so that exposure to the weather will not interfere with successful operation.

CHAPTER 3

GENERAL INSTALLATION REQUIREMENTS

INTRODUCTION

This chapter contains provisions from Article 110—Requirements for Electrical Installations. These provisions are essential to the installation of conductors and equipment. Article 110 provides fundamental safety requirements that apply to all types of electrical installations. While certain provisions apply to all aspects of electrical installations, others pertain to specific areas. For example, conductors and equipment required (or permitted) by the *Code* must be approved [110.2]. While this requirement applies to the installation of all conductors and equipment, other provisions are more specific. For example, the requirements of 110.26(F) apply only to switchboards, panelboards, distribution boards, and motor control centers.

ARTICLE 110
Requirements for Electrical Installations

I. GENERAL

110.1 Scope. This article covers general requirements for the examination and approval, installation and use, access to and spaces about electrical conductors and equipment, and tunnel installations.

110.2 Approval. The conductors and equipment required or permitted by this *Code* shall be acceptable only if approved.

110.3 Examination, Identification, Installation, and Use of Equipment.

(A) Examination. In judging equipment, considerations such as the following shall be evaluated:

(1) Suitability for installation and use in conformity with the provisions of this *Code*
(2) Mechanical strength and durability, including, for parts designed to enclose and protect other equipment, the adequacy of the protection thus provided
(3) Wire-bending and connection space
(4) Electrical insulation
(5) Heating effects under normal conditions of use and also under abnormal conditions likely to arise in service
(6) Arcing effects
(7) Classification by type, size, voltage, current capacity, and specific use
(8) Other factors that contribute to the practical safeguarding of persons using or likely to come in contact with the equipment

(B) Installation and Use. Listed or labeled equipment shall be installed and used in accordance with any instructions included in the listing or labeling.

110.5 Conductors. Conductors normally used to carry current shall be of copper unless otherwise provided in this *Code*. Where the conductor material is not specified, the material and the sizes given in this *Code* shall apply to copper conductors. Where other materials are used, the size shall be changed accordingly.

110.6 Conductor Sizes. Conductor sizes are expressed in American Wire Gage (AWG) or in circular mils.

110.7 Insulation Integrity. Completed wiring installations shall be free from short circuits and from grounds other than as required or permitted in Article 250.

110.11 Deteriorating Agents. Unless identified for use in the operating environment, no conductors or equipment shall be located in damp or wet locations; where exposed to gases, fumes, vapors, liquids, or other agents that have a deteriorating effect on the conductors or equipment; or where exposed to excessive temperatures.

Equipment identified only as "dry locations," "Type 1," or "indoor use only" shall be protected against permanent damage from the weather during building construction.

110.12 Mechanical Execution of Work. Electrical equipment shall be installed in a neat and workmanlike manner.

(A) Unused Openings. Unused cable or raceway openings in boxes, raceways, auxiliary gutters, cabinets, cutout boxes, meter socket enclosures, equipment cases, or housings shall be effectively closed to afford protection substantially equivalent to the wall of the equipment. Where metallic plugs or plates are used with nonmetallic enclosures, they shall be recessed at least 6 mm (1/4 in.) from the outer surface of the enclosure.

(C) Integrity of Electrical Equipment and Connections. Internal parts of electrical equipment, including busbars, wiring terminals, insulators, and other surfaces, shall not be damaged or contaminated by foreign materials such as paint, plaster, cleaners, abrasives, or corrosive residues. There shall be no damaged parts that may adversely affect safe operation or mechanical strength of the equipment such as parts that are broken; bent; cut; or deteriorated by corrosion, chemical action, or overheating.

110.13 Mounting and Cooling of Equipment.

(A) Mounting. Electrical equipment shall be firmly secured to the surface on which it is mounted. Wooden plugs driven into holes in masonry, concrete, plaster, or similar materials shall not be used.

110.14 Electrical Connections. Because of different characteristics of dissimilar metals, devices such as pressure terminal or pressure splicing connectors and soldering lugs shall be identified for the material of the conductor and shall be properly installed and used. Conductors of dissimilar metals shall not be intermixed in a terminal or splicing connector where physical contact occurs between dissimilar conductors (such as copper and aluminum, copper and copper-clad aluminum, or aluminum and copper-clad aluminum), unless the device is identified for the purpose and conditions of use. Materials such as solder, fluxes, inhibitors, and compounds, where employed, shall be suitable for the use and shall be of a type that will not adversely affect the conductors, installation, or equipment.

(A) Terminals. Connection of conductors to terminal parts shall ensure a thoroughly good connection without damaging the conductors and shall be made by means of pressure connectors (including set-screw type), solder lugs, or splices to flexible leads. Connection by means of wire-binding screws or studs and nuts that have upturned lugs or the equivalent shall be permitted for 10 AWG or smaller conductors.

Terminals for more than one conductor and terminals used to connect aluminum shall be so identified.

(B) Splices. Conductors shall be spliced or joined with splicing devices identified for the use or by brazing, welding, or soldering with a fusible metal or alloy. Soldered splices shall first be spliced or joined so as to be mechanically and electrically secure without solder and then be soldered. All splices and joints and the free ends of conductors shall be covered with an insulation equivalent to that of the conductors or with an insulating device identified for the purpose.

Wire connectors or splicing means installed on conductors for direct burial shall be listed for such use.

(C) Temperature Limitations. The temperature rating associated with the ampacity of a conductor shall be selected and coordinated so as not to exceed the lowest temperature rating of any connected termination, conductor, or device. Conductors with temperature ratings higher than specified for terminations shall be permitted to be used for ampacity adjustment, correction, or both.

(1) Equipment Provisions. The determination of termination provisions of equipment shall be based on 110.14(A) or (B). Unless the equipment is listed and marked otherwise, conductor ampacities used in determining equipment termination provisions shall be based on Table 310.16 as appropriately modified by 310.15(B)(1) through (6).

(a) Termination provisions of equipment for circuits rated 100 amperes or less, or marked for 14 AWG through 1 AWG conductors, shall be used only for one of the following:

(1) Conductors rated 60°C (140°F)
(2) Conductors with higher temperature ratings, provided the ampacity of such conductors is determined based on the 60°C (140°F) ampacity of the conductor size used
(3) Conductors with higher temperature ratings if the equipment is listed and identified for use with such conductors

(b) Termination provisions of equipment for circuits rated over 100 amperes, or marked for conductors larger than 1 AWG, shall be used only for one of the following:

(1) Conductors rated 75°C (167°F)
(2) Conductors with higher temperature ratings, provided the ampacity of such conductors does not exceed the 75°C (167°F) ampacity of the conductor size used, or up to their ampacity if the equipment is listed and identified for use with such conductors

110.22 Identification of Disconnecting Means. Each disconnecting means shall be legibly marked to indicate its purpose unless located and arranged so the purpose is evident. The marking shall be of sufficient durability to withstand the environment involved.

Where circuit breakers or fuses are applied in compliance with the series combination ratings marked on the equipment by the manufacturer, the equipment enclosure(s) shall be legibly marked in the field to indicate the equipment has been applied with a series combination rating.

The marking shall be readily visible and state the following:

CAUTION—SERIES COMBINATION SYSTEM
RATED ____ AMPERES. IDENTIFIED
REPLACEMENT COMPONENTS REQUIRED.

II. 600 VOLTS, NOMINAL, OR LESS

110.26 Spaces About Electrical Equipment. Sufficient access and working space shall be provided and maintained about all electric equipment to permit ready and safe operation and maintenance of such equipment. Enclosures housing electrical apparatus that are controlled by lock and key shall be considered accessible to qualified persons.

(A) Working Space. Working space for equipment operating at 600 volts, nominal, or less to ground and likely to require examination, adjustment, servicing, or maintenance while energized shall comply with the dimensions of 110.26(A)(1), (2), and (3) or as required or permitted elsewhere in this *Code*.

(1) Depth of Working Space. The depth of the working space in the direction of live parts shall not be less than that specified in Table 110.26(A)(1) unless the requirements of 110.26(A)(1)(a), (b), or (c) are met. Distances shall be measured from the exposed live parts or from the enclosure or opening if the live parts are enclosed.

(2) Width of Working Space. The width of the working space in front of the electric equipment shall be the width of the equipment or 750 mm (30 in.), whichever is greater. In all cases, the work space shall permit at least a 90 degree opening of equipment doors or hinged panels.

(3) Height of Working Space. The work space shall be clear and extend from the grade, floor, or platform to the height required by 110.26(E). Within the height require-

Table 110.26(A)(1) Working Spaces

Nominal Voltage to Ground	Minimum Clear Distance		
	Condition 1	Condition 2	Condition 3
0–150	900 mm (3 ft)	900 mm (3 ft)	900 mm (3 ft)
151–600	900 mm (3 ft)	1 m (3-1/2 ft)	1.2 m (4 ft)

Note: Where the conditions are as follows:

Condition 1—Exposed live parts on one side and no live or grounded parts on the other side of the working space, or exposed live parts on both sides effectively guarded by suitable wood or other insulating materials. Insulated wire or insulated busbars operating at not over 300 volts to ground shall not be considered live parts.

Condition 2—Exposed live parts on one side and grounded parts on the other side. Concrete, brick, or tile walls shall be considered as grounded.

Condition 3—Exposed live parts on both sides of the work space (not guarded as provided in Condition 1) with the operator between.

ments of this section, other equipment that is associated with the electrical installation and is located above or below the electrical equipment shall be permitted to extend not more than 150 mm (6 in.) beyond the front of the electrical equipment.

(B) Clear Spaces. Working space required by this section shall not be used for storage. When normally enclosed live parts are exposed for inspection or servicing, the working space, if in a passageway or general open space, shall be suitably guarded.

(D) Illumination. Illumination shall be provided for all working spaces about service equipment, switchboards, panelboards, or motor control centers installed indoors. Additional lighting outlets shall not be required where the work space is illuminated by an adjacent light source or as permitted by 210.70(A)(1), Exception No. 1, for switched receptacles. In electrical equipment rooms, the illumination shall not be controlled by automatic means only.

(E) Headroom. The minimum headroom of working spaces about service equipment, switchboards, panelboards, or motor control centers shall be 2.0 m (6-1/2 ft). Where the electrical equipment exceeds 2.0 m (6-1/2 ft) in height, the minimum headroom shall not be less than the height of the equipment.

Exception: In existing dwelling units, service equipment or panelboards that do not exceed 200 amperes shall be permitted in spaces where the headroom is less than 2.0 (6-1/2 ft).

(F) Dedicated Equipment Space. All switchboards, panelboards, distribution boards, and motor control centers, shall be located in dedicated spaces and protected from damage.

Exception: Control equipment that by its very nature or because of other rules of the Code must be adjacent to or within sight of its operating machinery shall be permitted in those locations.

(1) Indoor. Indoor installations shall comply with 110.26(F)(1)(a) through (d).

(a) Dedicated Electrical Space. The space equal to the width and depth of the equipment and extending from the floor to a height of 1.8 m (6 ft) above the equipment or to the structural ceiling, whichever is lower, shall be dedicated to the electrical installation. No piping, ducts, leak protection apparatus, or other equipment foreign to the electrical installation shall be located in this zone.

Exception: Suspended ceilings with removable panels shall be permitted within the 1.8-m (6-ft) zone.

(b) Foreign Systems. The area above the dedicated space required by 110.26(F)(1)(a) shall be permitted to contain foreign systems, provided protection is installed to avoid damage to the electrical equipment from condensation, leaks, or breaks in such foreign systems.

(d) Suspended Ceilings. A dropped, suspended, or similar ceiling that does not add strength to the building structure shall not be considered a structural ceiling.

(2) Outdoor. Outdoor electrical equipment shall be installed in suitable enclosures and shall be protected from accidental contact by unauthorized personnel, or by vehicular traffic, or by accidental spillage or leakage from piping systems. The working clearance space shall include the zone described in 110.26(A). No architectural appurtenance or other equipment shall be located in this zone.

110.27 Guarding of Live Parts.

(B) Prevent Physical Damage. In locations where electric equipment is likely to be exposed to physical damage, enclosures or guards shall be so arranged and of such strength as to prevent such damage.

CHAPTER 4

TEMPORARY ELECTRICAL POWER

INTRODUCTION

This chapter contains provisions from Article 527—Temporary Installations. Article 527 covers provisions pertaining to temporary electrical power and lighting. New construction generally requires the installation of temporary electrical equipment, which usually involves setting a pole and building a service. Although the feeders, branch circuits, and receptacles are specifically covered by Article 527, the service must be installed in accordance with Article 230 (see Chapter 6 of this book). For example, the minimum clearance of overhead service-drop conductors must comply with 230.24. Of course, not all electrical projects will require the installation of a temporary service. Remodeling and additions to existing installations typically already have available power through permanent wiring.

Because of the inherent hazards associated with construction sites, ground-fault protection of receptacles with certain ratings that are in use by personnel is required for all temporary wiring installations, not just for new construction sites with temporary services. If a receptacle(s) is installed or exists as part of the permanent wiring or the building or structure and is used for temporary electric power, ground-fault circuit-interrupter (GFCI) protection for personnel must be provided. This may require replacing the existing receptacle with a GFCI receptacle or the use of a portable GFCI device. All other outlets (besides 125-volt, single-phase, 15-, 20-, and 30-ampere receptacle outlets) must have GFCI protection for personnel, or an assured equipment grounding conductor program in accordance with 527.6(B)(2).

ARTICLE 527
Temporary Installations

527.1 Scope. The provisions of this article apply to temporary electrical power and lighting installations.

527.2 All Wiring Installations.

(A) Other Articles. Except as specifically modified in this article, all other requirements of this *Code* for permanent wiring shall apply to temporary wiring installations.

(B) Approval. Temporary wiring methods shall be acceptable only if approved based on the conditions of use and any special requirements of the temporary installation.

527.3 Time Constraints.

(A) During the Period of Construction. Temporary electrical power and lighting installations shall be permitted during the period of construction, remodeling, maintenance, repair, or demolition of buildings, structures, equipment, or similar activities.

(D) Removal. Temporary wiring shall be removed immediately upon completion of construction or purpose for which the wiring was installed.

527.4 General.

(A) Services. Services shall be installed in conformance with Article 230.

(B) Feeders. Feeders shall be protected as provided in Article 240. They shall originate in an approved distribution center. Conductors shall be permitted within cable assemblies or within multiconductor cords or cables of a type identified in Table 400.4 for hard usage or extra-hard usage. For the purpose of this section, Type NM and Type NMC cables shall be permitted to be used in any dwelling, building, or structure without any height limitation.

(C) Branch Circuits. All branch circuits shall originate in an approved power outlet or panelboard. Conductors shall be permitted within cable assemblies or within multiconductor cord or cable of a type identified in Table 400.4 for hard usage or extra-hard usage. All conductors shall be protected as provided in Article 240. For the purposes of this section, Type NM and Type NMC cables shall be permitted to be used in any dwelling, building, or structure without any height limitation.

(D) Receptacles. All receptacles shall be of the grounding type. Unless installed in a continuous grounded metal raceway or metal-covered cable, all branch circuits shall contain a separate equipment grounding conductor, and all receptacles shall be electrically connected to the equipment grounding conductors. Receptacles on construction sites shall not be installed on branch circuits that supply temporary lighting. Receptacles shall not be connected to the same ungrounded conductor of multiwire circuits that supply temporary lighting.

(E) Disconnecting Means. Suitable disconnecting switches or plug connectors shall be installed to permit the disconnection of all ungrounded conductors of each temporary circuit. Multiwire branch circuits shall be provided with a means to disconnect simultaneously all ungrounded conductors at the power outlet or panelboard where the branch circuit originated. Approved handle ties shall be permitted.

(F) Lamp Protection. All lamps for general illumination shall be protected from accidental contact or breakage by a suitable fixture or lampholder with a guard.

Brass shell, paper-lined sockets, or other metal-cased sockets shall not be used unless the shell is grounded.

(G) Splices. On construction sites, a box shall not be required for splices or junction connections where the circuit conductors are multiconductor cord or cable assemblies, provided that the equipment grounding continuity is maintained with or without the box. See 110.14(B) and 400.9. A box, conduit body, or terminal fitting having a separately bushed hole for each conductor shall be used wherever a change is made to a conduit or tubing system or a metal-sheathed cable system.

(H) Protection from Accidental Damage. Flexible cords and cables shall be protected from accidental damage. Sharp corners and projections shall be avoided. Where passing through doorways or other pinch points, protection shall be provided to avoid damage.

(I) Termination(s) at Devices. Flexible cords and cables entering enclosures containing devices requiring termination shall be secured to the box with fittings designed for the purpose.

(J) Support. Cable assemblies and flexible cords and cables shall be supported in place at intervals that ensure that they will be protected from physical damage. Support shall be in the form of staples, cable ties, straps, or similar type fittings installed so as not to cause damage. Vegetation shall not be used for support of overhead spans of branch circuits or feeders.

527.6 Ground-Fault Protection for Personnel. Ground-fault protection for personnel for all temporary wiring installations shall be provided to comply with 527.6(A) and (B). This section shall apply only to temporary wiring installations used to supply temporary power to equipment used by personnel during construction, remodeling, maintenance, repair, or demolition of buildings, structures, equipment, or similar activities.

(A) Receptacle Outlets. All 125-volt, single-phase, 15-, 20-, and 30-ampere receptacle outlets that are not a part of the permanent wiring of the building or structure and that are in use by personnel shall have ground-fault circuit interrupter protection for personnel. If a receptacle(s) is installed or exists as part of the permanent wiring of the building or structure and is used for temporary electric power, ground-fault circuit-interrupter protection for personnel shall be provided. For the purposes of this section, cord sets or devices incorporating listed ground-fault circuit interrupter protection for personnel identified for portable use shall be permitted.

(B) Use of Other Outlets. Receptacles other than 125-volt, single-phase, 15-, 20-, and 30-ampere receptacles shall have protection in accordance with (1) or, the assured equipment grounding conductor program in accordance with (2).

(1) GFCI Protection. Ground-fault circuit interrupter protection for personnel.

CHAPTER 5

SERVICE AND FEEDER LOAD CALCULATIONS

INTRODUCTION

Selecting the size service and panelboard(s) is one of the first steps in residential installations. Load calculations are required to size the service and/or feeder overcurrent protection, as well as the ungrounded, grounded, and grounding conductors. Article 220 contains detailed requirements for computing branch-circuit, feeder, and service loads for one-family, two-family, and multifamily dwellings. Two load calculation methods are provided, *Standard* (Part II) and *Optional* (Part III).

While certain installations will have a service with a single panelboard, other installations may have multiple panelboards. For dwellings with panelboards that are not part of the service, additional load calculations are necessary—one for the service and one for each feeder.

Collecting specific information and applying the appropriate demand factors will provide answers to such questions as: What is the minimum size service or feeder? What is the minimum size ungrounded (hot) conductor? What is the minimum size neutral conductor? What is the minimum size grounding electrode conductor (for services) or equipment grounding conductor (for feeders)?

ARTICLE 220
Branch-Circuit, Feeder, and Service Calculations

I. GENERAL

220.1 Scope. This article provides requirements for computing branch-circuit, feeder, and service loads.

220.2 Computations.

(A) Voltages. Unless other voltages are specified, for purposes of computing branch-circuit and feeder loads, nominal system voltages of 120, 120/240, 208Y/120, 240, 347, 480Y/277, 480, 600Y/347, and 600 volts shall be used.

(B) Fractions of an Ampere. Where computations result in a fraction of an ampere that is less than 0.5, such fractions shall be permitted to be dropped.

220.3 Computation of Branch Circuit Loads. Branch-circuit loads shall be computed as shown in 220.3(A) through (C).

(A) Lighting Load for Specified Occupancies. A unit load of not less than that specified in Table 220.3(A) for occupancies specified therein shall constitute the minimum lighting load. The floor area for each floor shall be computed from the outside dimensions of the building, dwelling unit, or other area involved. For dwelling units, the computed floor area shall not include open porches, garages, or unused or unfinished spaces not adaptable for future use.

Note…residential installations require a unit load of 3 volt-amperes per square foot (or 33 volt-amperes per square meter.)

(B) Other Loads—All Occupancies. In all occupancies, the minimum load for each outlet for general-use receptacles and outlets not used for general illumination shall not be less than that computed in 220.3(B)(1) through (11), the loads shown being based on nominal branch-circuit voltages.

(1) Specific Appliances or Loads. An outlet for a specific appliance or other load not covered in (2) through (11) shall be computed based on the ampere rating of the appliance or load served.

(2) Electric Dryers and Household Electric Cooking Appliances. Load computations shall be permitted as specified in 220.18 for electric dryers and in 220.19 for electric ranges and other cooking appliances.

(3) Motor Loads. Outlets for motor loads shall be computed in accordance with the requirements in 430.22, 430.24, and 440.6.

(4) Recessed Luminaires (Lighting Fixtures). An outlet supplying recessed luminaire(s) [lighting fixture(s)] shall be computed based on the maximum volt-ampere rating of the equipment and lamps for which the luminaire(s) [fixture(s)] is rated.

(10) Dwelling Occupancies. In one-family, two-family, and multifamily dwellings and in guest rooms of hotels and motels, the outlets specified in (1), (2), and (3) are included in the general lighting load calculations of 220.3(A). No additional load calculations shall be required for such outlets.

(1) All general-use receptacle outlets of 20-ampere rating or less, including receptacles connected to the circuits in 210.11(C)(3)
(2) The receptacle outlets specified in 210.52(E) and (G)
(3) The lighting outlets specified in 210.70(A) and (B)

II. FEEDERS AND SERVICES

220.10 General. The computed load of a feeder or service shall not be less than the sum of the loads on the branch circuits supplied, as determined by Part I of this article, after any applicable demand factors permitted by Parts II, III, or IV have been applied.

220.11 General Lighting. The demand factors specified in Table 220.11 shall apply to that portion of the total branch-circuit load computed for general illumination. They shall not be applied in determining the number of branch circuits for general illumination.

220.14 Motors. Motor loads shall be computed in accordance with 430.24, 430.25, and 430.26 and with 440.6 for hermetic refrigerant motor compressors.

Table 220.11 Lighting Load Demand Factors

Type of Occupancy	Portion of Lighting Load to Which Demand Factor Applies (Volt-Amperes)	Demand Factor (Percent)
Dwelling units	First 3000 or less at	100
	From 3001 to 120,000 at	35
	Remainder over 120,000 at	25

220.15 Fixed Electric Space Heating. Fixed electric space heating loads shall be computed at 100 percent of the total connected load; however, in no case shall a feeder or service load current rating be less than the rating of the largest branch circuit supplied.

220.16 Small Appliance and Laundry Loads—Dwelling Unit.

(A) Small Appliance Circuit Load. In each dwelling unit, the load shall be computed at 1500 volt-amperes for each 2-wire small-appliance branch circuit required by 210.11(C)(1). Where the load is subdivided through two or more feeders, the computed load for each shall include not less than 1500 volt-amperes for each 2-wire small-appliance branch circuit. These loads shall be permitted to be included with the general lighting load and subjected to the demand factors provided in Table 220.11.

Exception: The individual branch circuit permitted by 210.52(B)(1), Exception No. 2, shall be permitted to be excluded from the calculation required by 220.16.

(B) Laundry Circuit Load. A load of not less than 1500 volt-amperes shall be included for each 2-wire laundry branch circuit installed as required by 210.11(C)(2). This load shall be permitted to be included with the general lighting load and subjected to the demand factors provided in Table 220.11.

220.17 Appliance Load—Dwelling Unit(s). It shall be permissible to apply a demand factor of 75 percent to the nameplate rating load of four or more appliances fastened in place, other than electric ranges, clothes dryers, space-heating equipment, or air-conditioning equipment, that are served by the same feeder or service in a one-family, two-family, or multifamily dwelling.

220.18 Electric Clothes Dryers—Dwelling Unit(s). The load for household electric clothes dryers in a dwelling unit(s) shall be 5000 watts (volt-amperes) or the nameplate rating, whichever is larger, for each dryer served. The use of the demand factors in Table 220.18 shall be permitted. Where two or more single-phase dryers are supplied by a 3-phase, 4-wire feeder or service, the total load shall be computed on the basis of twice the maximum number connected between any two phases.

220.19 Electric Ranges and Other Cooking Appliances—Dwelling Unit(s). The demand load for household electric ranges, wall-mounted ovens, counter-mounted cooking units, and other household cooking appliances individually rated in excess of 1-3/4 kW shall be permitted to be computed in accordance with Table 220.19. Kilovolt-amperes (kVA) shall be

**Table 220.18 Demand Factors for Household
Electric Clothes Dryers**

Number of Dryers	Demand Factor (Percent)
1–4	100%
5	85%
6	75%
7	65%
8	60%
9	55%
10	50%
11	47%
12–22	% = 47 – (number of dryers – 11)
23	35%
24–42	% = 35 – [0.5 × (number of dryers – 23)]
43 and over	25%

Table 220.19 Demand Loads for Household Electric Ranges, Wall-Mounted Ovens, Counter-Mounted Cooking Units, and Other Household Cooking Appliances over 1-3/4 kW Rating (Column C to be used in all cases except as otherwise permitted in Note 3.)

Number of Appliances	Demand Factor (Percent) (See Notes)		Column C Maximum Demand (kW) (See Notes) (Not over 12 kW Rating)
	Column A (Less than 3-1/2 kW Rating)	Column B (3-1/2 kW to 8-3/4 kW Rating)	
1	80	80	8
2	75	65	11
3	70	55	14
4	66	50	17
5	62	45	20
6	59	43	21
7	56	40	23
8	53	36	23
9	51	35	24
10	49	34	25
11	47	32	26
12	45	32	27

13	43	32	28
14	41	32	29
15	40	32	30
16	39	28	31
17	38	28	32
18	37	28	33
19	36	28	34
20	35	28	35
21	34	26	36
22	33	26	37
23	32	26	38
24	31	26	39
25	30	26	40
26–30	30	24	15 kW + 1 kW
31–40	30	22	for each range
41–50	30	20	25 kW + 3/4 kW
51–60	30	18	for each range
61 and over	30	16	

Continued

Table 220.19 Demand Loads for Household Electric Ranges, Wall-Mounted Ovens, Counter-Mounted Cooking Units, and Other Household Cooking Appliances over 1-3/4 kW Rating (Column C to be used in all cases except as otherwise permitted in Note 3.) *Continued*

1. Over 12 kW through 27 kW ranges all of same rating. For ranges individually rated more than 12 kW but not more than 27 kW, the maximum demand in Column C shall be increased 5 percent for each additional kilowatt of rating or major fraction thereof by which the rating of individual ranges exceeds 12 kW.

2. Over 8-3/4 kW through 27 kW ranges of unequal ratings. For ranges individually rated more than 8-3/4 kW and of different ratings, but none exceeding 27 kW, an average value of rating shall be computed by adding together the ratings of all ranges to obtain the total connected load (using 12 kW for any range rated less than 12 kW) and dividing by the total number of ranges. Then the maximum demand in Column C shall be increased 5 percent for each kilowatt or major fraction thereof by which this average value exceeds 12 kW.

3. Over 1-3/4 kW through 8-3/4 kW. In lieu of the method provided in Column C, it shall be permissible to add the nameplate ratings of all household cooking appliances rated more than 1-3/4 kW but not more than 8-3/4 kW and multiply the sum by the demand factors specified in Column A or B for the given number of appliances. Where the rating of cooking appliances falls under both Column A and Column B, the demand factors for each column shall be applied to the appliances for that column, and the results added together.

4. Branch-Circuit Load. It shall be permissible to compute the branch-circuit load for one range in accordance with Table 220.19. The branch-circuit load for one wall-mounted oven or one counter-mounted cooking unit shall be the nameplate rating of the appliance. The branch-circuit load for a counter-mounted cooking unit and not more than two wall-mounted ovens, all supplied from a single branch circuit and located in the same room, shall be computed by adding the nameplate rating of the individual appliances and treating this total as equivalent to one range.

considered equivalent to kilowatts (kW) for loads computed under this section.

Where two or more single-phase ranges are supplied by a 3-phase, 4-wire feeder or service, the total load shall be computed on the basis of twice the maximum number connected between any two phases.

220.21 Noncoincident Loads. Where it is unlikely that two or more noncoincident loads will be in use simultaneously, it shall be permissible to use only the largest load(s) that will be used at one time, in computing the total load of a feeder or service.

220.22 Feeder or Service Neutral Load. The feeder or service neutral load shall be the maximum unbalance of the load determined by this article. The maximum unbalanced load shall be the maximum net computed load between the neutral and any one ungrounded conductor, except that the load thus obtained shall be multiplied by 140 percent for 3-wire, 2-phase or 5-wire, 2-phase systems. For a feeder or service supplying household electric ranges, wall-mounted ovens, counter-mounted cooking units, and electric dryers, the maximum unbalanced load shall be considered as 70 percent of the load on the ungrounded conductors, as determined in accordance with Table 220.19 for ranges and Table 220.18 for dryers. For 3-wire dc or single-phase ac; 4-wire, 3-phase; 3-wire, 2-phase; or 5-wire, 2-phase systems, a further demand factor of 70 percent shall be permitted for that portion of the unbalanced load in excess of 200 amperes. There shall be no reduction of the neutral capacity for that portion of the load that consists of nonlinear loads supplied from a 4-wire, wye-connected, 3-phase system. There shall be no reduction in the capacity of the grounded conductor of a 3-wire circuit consisting of two phase wires and the neutral of a 4-wire, 3-phase, wye-connected system.

III. OPTIONAL CALCULATIONS FOR COMPUTING FEEDER AND SERVICE LOADS

220.30 Optional Calculation—Dwelling Unit.

(A) Feeder and Service Load. For a dwelling unit having the total connected load served by a single 3-wire,

120/240-volt or 208Y/120-volt set of service or feeder conductors with an ampacity of 100 or greater, it shall be permissible to compute the feeder and service loads in accordance with this section instead of the method specified in Part II of this article. The calculated load shall be the result of adding the loads from 220.30(B) and (C). Feeder and service-entrance conductors whose demand load is determined by this optional calculation shall be permitted to have the neutral load determined by 220.22.

(B) General Loads. The general calculated load shall be not less than 100 percent of the first 10 kVA plus 40 percent of the remainder of the following loads:

(1) 1500 volt-amperes for each 2-wire, 20-ampere small-appliance branch circuit and each laundry branch circuit specified in 220.16.
(2) 33 volt-amperes/m^2 or 3 volt-amperes/ft^2 for general lighting and general-use receptacles. The floor area for each floor shall be computed from the outside dimensions of the dwelling unit. The computed floor area shall not include open porches, garages, or unused or unfinished spaces not adaptable for future use.
(3) The nameplate rating of all appliances that are fastened in place, permanently connected, or located to be on a specific circuit, ranges, wall-mounted ovens, counter-mounted cooking units, clothes dryers, and water heaters.
(4) The nameplate ampere or kVA rating of all motors and of all low-power-factor loads.

(C) Heating and Air-Conditioning Load. The largest of the following six selections (load in kVA) shall be included:

(1) 100 percent of the nameplate rating(s) of the air conditioning and cooling.
(2) 100 percent of the nameplate ratings of the heat pump compressors and supplemental heating unless the controller prevents the compressor and supplemental heating from operating at the same time.
(3) 100 percent of the nameplate ratings of electric thermal storage and other heating systems where the usual load is expected to be continuous at the full nameplate

value. Systems qualifying under this selection shall not be calculated under any other selection in 220.30(C).

(4) 65 percent of the nameplate rating(s) of the central electric space heating, including integral supplemental heating in heat pumps where the controller prevents the compressor and supplemental heating from operating at the same time.

(5) 65 percent of the nameplate rating(s) of electric space heating if less than four separately controlled units.

(6) 40 percent of the nameplate rating(s) of electric space heating if four or more separately controlled units.

220.31 Optional Calculations for Additional Loads in an Existing Dwelling Unit. This section shall be permitted to be used to determine if the existing service or feeder is of sufficient capacity to serve additional loads. Where the dwelling unit is served by a 120/240-volt or 208Y/120-volt, 3-wire service, it shall be permissible to compute the total load in accordance with 220.31(A) or (B).

(A) Where Additional Air-Conditioning Equipment or Electric Space-Heating Equipment Is Not to Be Installed. The following formula shall be used for existing and additional new loads.

Load (kVA)	Percent of Load
First 8 kVa of load at	100
Remainder of load at	40

Load calculations shall include the following:

(1) General lighting and general-use receptacles at 33 volt-amperes/m^2 or 3 volt-amperes/ft^2 as determined by 220.3(A)

(2) 1500 volt-amperes for each 2-wire, 20-ampere small-appliance branch circuit and each laundry branch circuit specified in 220.16

(3) Household range(s), wall-mounted oven(s), and counter-mounted cooking unit(s)

(4) All other appliances that are permanently connected, fastened in place, or connected to a dedicated circuit, at nameplate rating

(B) Where Additional Air-Conditioning Equipment or Electric Space-Heating Equipment Is to Be Installed. The following formula shall be used for existing and additional new loads. The larger connected load of air-conditioning or space-heating, but not both, shall be used.

Air-conditioning equipment	100
Central electric space heating	100
Less than four separately controlled space-heating units	100
First 8 kVA of all other loads	100
Remainder of all other loads	40

Other loads shall include the following:

(1) General lighting and general-use receptacles at 33 volt-amperes/m² or 3 volt-amperes/ft² as determined by 220.3(A)
(2) 1500 volt-amperes for each 2-wire, 20-ampere small-appliance branch circuit and each laundry branch circuit specified in 220.16
(3) Household range(s), wall-mounted oven(s), and counter-mounted cooking unit(s)
(4) All other appliances that are permanently connected, fastened in place, or connected to a dedicated circuit, including four or more separately controlled space-heating units, at nameplate rating

220.32 Optional Calculation—Multifamily Dwelling.

(A) Feeder or Service Load. It shall be permissible to compute the load of a feeder or service that supplies more than two dwelling units of a multifamily dwelling in accordance with Table 220.32 instead of Part II of this article where all the following conditions are met:

(1) No dwelling unit is supplied by more than one feeder.
(2) Each dwelling unit is equipped with electric cooking equipment.
(3) Each dwelling unit is equipped with either electric space heating, air conditioning, or both. Feeders and service conductors whose demand load is determined by this optional calculation shall be permitted to have the neutral load determined by 220.22.

(B) House Loads. House loads shall be computed in accordance with Part II of this article and shall be in addition to the dwelling unit loads computed in accordance with Table 220.32.

(C) Connected Loads. The computed load to which the demand factors of Table 220.32 apply shall include the following:

(1) 1500 volt-amperes for each 2-wire, 20-ampere small-appliance branch circuit and each laundry branch circuit specified in 220.16.
(2) 33 volt-amperes/m^2 or 3 volt-amperes/ft^2 for general lighting and general-use receptacles.

Table 220.32 Optional Calculations—Demand Factors for Three or More Multifamily Dwelling Units

Number of Dwelling Units	Demand Factor (Percent)
3–5	45
6–7	44
8–10	43
11	42
12–13	41
14–15	40
16–17	39
18–20	38
21	37
22–23	36
24–25	35
26–27	34
28–30	33
31	32
32–33	31
34–36	30
37–38	29
39–42	28
43–45	27
46–50	26
51–55	25
56–61	24
62 and over	23

(3) The nameplate rating of all appliances that are fastened in place, permanently connected or located to be on a specific circuit, ranges, wall-mounted ovens, counter-mounted cooking units, clothes dryers, water heaters, and space heaters. If water heater elements are interlocked so that all elements cannot be used at the same time, the maximum possible load shall be considered the nameplate load.

(4) The nameplate ampere or kilovolt-ampere rating of all motors and of all low-power-factor loads.

(5) The larger of the air-conditioning load or the space-heating load.

220.33 Optional Calculation—Two Dwelling Units. Where two dwelling units are supplied by a single feeder and the computed load under Part II of this article exceeds that for three identical units computed under 220.32, the lesser of the two loads shall be permitted to be used.

CHAPTER 6

SERVICES

INTRODUCTION

This chapter contains provisions from Article 230—Services. Article 230 covers service conductors and equipment for control and protection of services and their installation requirements. Contained in this pocket guide are seven of the eight parts that make up Article 230. The seven parts are: I General; II Overhead Service-Drop Conductors; III Underground Service-Lateral Conductors; IV Service-Entrance Conductors; V Service Equipment—General; VI Service Equipment—Disconnecting Means; and VII Service Equipment—Overcurrent Protection. Reference the *2002 National Electrical Code* for services exceeding 600 volts, nominal.

While the service installation must comply with Parts I, IV, V, VI, and VII, compliance with Parts II and III depends on the type of service entrance. Part II contains requirements for overhead service-drop conductors, and Part III contains requirements for underground service-lateral conductors.

ARTICLE 230
Services

230.1 Scope. This article covers service conductors and equipment for control and protection of services and their installation requirements.

FPN: See Figure 230.1.

I. GENERAL

230.2 Number of Services. A building or other structure served shall be supplied by only one service unless permitted in 230.2(A) through (D). For the purpose of 230.40, Exception No. 2 only, underground sets of conductors, 1/0 AWG and larger, running to the same location and connected together at their supply end but not connected together at their load end shall be considered to be supplying one service.

(A) Special Conditions. Additional services shall be permitted to supply the following:

(1) Fire pumps
(2) Emergency systems
(3) Legally required standby systems
(4) Optional standby systems
(5) Parallel power production systems

(B) Special Occupancies. By special permission, additional services shall be permitted for the following:

(1) Multiple-occupancy buildings where there is no available space for service equipment accessible to all occupants, or
(2) A single building or other structure sufficiently large to make two or more services necessary

(C) Capacity Requirements. Additional services shall be permitted under any of the following:

(1) Where the capacity requirements are in excess of 2000 amperes at a supply voltage of 600 volts or less
(2) Where the load requirements of a single-phase installation are greater than the serving agency normally supplies through one service
(3) By special permission

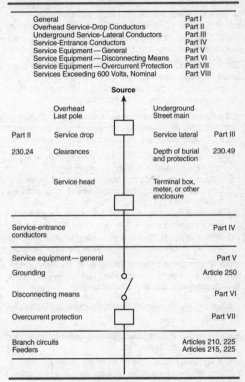

	General		Part I	
	Overhead Service-Drop Conductors		Part II	
	Underground Service-Lateral Conductors		Part III	
	Service-Entrance Conductors		Part IV	
	Service Equipment — General		Part V	
	Service Equipment — Disconnecting Means		Part VI	
	Service Equipment — Overcurrent Protection		Part VII	
	Services Exceeding 600 Volts, Nominal		Part VIII	

Source

	Overhead Last pole		Underground Street main	
Part II	Service drop		Service lateral	Part III
230.24	Clearances		Depth of burial and protection	230.49
	Service head		Terminal box, meter, or other enclosure	

Service-entrance conductors	Part IV
Service equipment — general	Part V
Grounding	Article 250
Disconnecting means	Part VI
Overcurrent protection	Part VII
Branch circuits Feeders	Articles 210, 225 Articles 215, 225

Figure 230.1 Services.

(D) Different Characteristics. Additional services shall be permitted for different voltages, frequencies, or phases, or for different uses, such as for different rate schedules.

(E) Identification. Where a building or structure is supplied by more than one service, or any combination of branch circuits, feeders, and services, a permanent plaque or directory shall be installed at each service disconnect location denoting all other services, feeders, and branch circuits supplying that building or structure and the area served by each. See 225.37.

230.3 One Building or Other Structure Not to Be Supplied Through Another. Service conductors supplying a building or other structure shall not pass through the interior of another building or other structure.

230.6 Conductors Considered Outside the Building. Conductors shall be considered outside of a building or other structure under any of the following conditions:

(1) Where installed under not less than 50 mm (2 in.) of concrete beneath a building or other structure
(2) Where installed within a building or other structure in a raceway that is encased in concrete or brick not less than 50 mm (2 in.) thick
(3) Where installed in any vault that meets the construction requirements of Article 450, Part III
(4) Where installed in conduit and under not less than 450 mm (18 in.) of earth beneath a building or other structure

230.7 Other Conductors in Raceway or Cable. Conductors other than service conductors shall not be installed in the same service raceway or service cable.

Exception No. 1: Grounding conductors and bonding jumpers.

Exception No. 2: Load management control conductors having overcurrent protection.

230.8 Raceway Seal. Where a service raceway enters a building or structure from an underground distribution system, it shall be sealed in accordance with 300.5(G). Spare or unused raceways shall also be sealed. Sealants shall be identified for use with the cable insulation, shield, or other components.

230.9 Clearance from Building Openings. Service conductors and final spans shall comply with 230.9(A), (B), and (C).

(A) Clearance from Windows. Service conductors installed as open conductors or multiconductor cable without an overall outer jacket shall have a clearance of not less than 900 mm (3 ft) from windows that are designed to be opened, doors, porches, balconies, ladders, stairs, fire escapes, or similar locations.

Exception: Conductors run above the top level of a window shall be permitted to be less than the 900-mm (3-ft) requirement.

(B) Vertical Clearance. The vertical clearance of final spans above, or within 900 mm (3 ft) measured horizontally of, platforms, projections, or surfaces from which they might be reached shall be maintained in accordance with 230.24(B).

(C) Building Openings. Overhead service conductors shall not be installed beneath openings through which materials may be moved, such as openings in farm and commercial buildings, and shall not be installed where they obstruct entrance to these building openings.

230.10 Vegetation as Support. Vegetation such as trees shall not be used for support of overhead service conductors.

II. OVERHEAD SERVICE-DROP CONDUCTORS

230.22 Insulation or Covering. Individual conductors shall be insulated or covered.

Exception: The grounded conductor of a multiconductor cable shall be permitted to be bare.

230.23 Size and Rating.

(A) General. Conductors shall have sufficient ampacity to carry the current for the load as computed in accordance with Article 220 and shall have adequate mechanical strength.

(B) Minimum Size. The conductors shall not be smaller than 8 AWG copper or 6 AWG aluminum or copper-clad aluminum.

(C) Grounded Conductors. The grounded conductor shall not be less than the minimum size as required by 250.24(B).

230.24 Clearances. Service-drop conductors shall not be readily accessible and shall comply with 230.24(A) through (D) for services not over 600 volts, nominal.

(A) Above Roofs. Conductors shall have a vertical clearance of not less than 2.5 m (8 ft) above the roof surface. The vertical clearance above the roof level shall be maintained for a distance of not less than 900 mm (3 ft) in all directions from the edge of the roof.

Exception No. 1: The area above a roof surface subject to pedestrian or vehicular traffic shall have a verticle clearance from the roof surface in accordance with the clearance requirements of 230.24(B).

Exception No. 2: Where the voltage between conductors does not exceed 300 and the roof has a slope of 100 mm (4 in.) in 300 mm (12 in.), or greater, a reduction in clearance to 900 mm (3 ft) shall be permitted.

Exception No. 3: Where the voltage between conductors does not exceed 300, a reduction in clearance above only the overhanging portion of the roof to not less than 450 mm (18 in.) shall be permitted if (1) not more than 1.8 m (6 ft) of service-drop conductors, 1.2 m (4 ft) horizontally, pass above the roof overhang, and (2) they are terminated at a through-the-roof raceway or approved support.

Exception No. 4: The requirement for maintaining the vertical clearance 900 mm (3 ft) from the edge of the roof shall not apply to the final conductor span where the service drop is attached to the side of a building.

(B) Vertical Clearance from Ground. Service-drop conductors, where not in excess of 600 volts, nominal, shall have the following minimum clearance from final grade:

(1) 3.0 m (10 ft)—at the electric service entrance to buildings, also at the lowest point of the drip loop of the building electric entrance, and above areas or sidewalks accessible only to pedestrians, measured from final grade or other accessible surface only for service-drop cables supported on and cabled together with a grounded bare messenger where the voltage does not exceed 150 volts to ground

(2) 3.7 m (12 ft)—over residential property and driveways, and those commercial areas not subject to truck traffic where the voltage does not exceed 300 volts to ground

(3) 4.5 m (15 ft)—for those areas listed in the 3.7 m (12 ft) classification where the voltage exceeds 300 volts to ground

(4) 5.5 m (18 ft)—over public streets, alleys, roads, parking areas subject to truck traffic, driveways on other than residential property, and other land such as cultivated, grazing, forest, and orchard

230.26 Point of Attachment. The point of attachment of the service-drop conductors to a building or other structure shall provide the minimum clearances as specified in 230.24. In no case shall this point of attachment be less than 3.0 m (10 ft) above finished grade.

230.27 Means of Attachment. Multiconductor cables used for service drops shall be attached to buildings or other structures by fittings identified for use with service conductors. Open conductors shall be attached to fittings identified for use with service conductors or to noncombustible, nonabsorbent insulators securely attached to the building or other structure.

230.28 Service Masts as Supports. Where a service mast is used for the support of service-drop conductors, it shall be of adequate strength or be supported by braces or guys to withstand safely the strain imposed by the service drop. Where raceway-type service masts are used, all raceway fittings shall be identified for use with service masts. Only power service-drop conductors shall be permitted to be attached to a service mast.

230.29 Supports over Buildings. Service-drop conductors passing over a roof shall be securely supported by substantial structures. Where practicable, such supports shall be independent of the building.

III. UNDERGROUND SERVICE-LATERAL CONDUCTORS

230.30 Insulation. Service-lateral conductors shall be insulated for the applied voltage.

Exception: A grounded conductor shall be permitted to be uninsulated as follows:

(a) *Bare copper used in a raceway.*
(b) *Bare copper for direct burial where bare copper is judged to be suitable for the soil conditions.*

(c) Bare copper for direct burial without regard to soil conditions where part of a cable assembly identified for underground use.

(d) Aluminum or copper-clad aluminum without individual insulation or covering where part of a cable assembly identified for underground use in a raceway or for direct burial.

230.31 Size and Rating.

(A) General. Service-lateral conductors shall have sufficient ampacity to carry the current for the load as computed in accordance with Article 220 and shall have adequate mechanical strength.

(B) Minimum Size. The conductors shall not be smaller than 8 AWG copper or 6 AWG aluminum or copper-clad aluminum.

(C) Grounded Conductors. The grounded conductor shall not be less than the minimum size required by 250.24(B).

230.32 Protection Against Damage. Underground service-lateral conductors shall be protected against damage in accordance with 300.5. Service-lateral conductors entering a building shall be installed in accordance with 230.6 or protected by a raceway wiring method identified in 230.43.

230.33 Spliced Conductors. Service-lateral conductors shall be permitted to be spliced or tapped in accordance with 110.14, 300.5(E), 300.13, and 300.15.

IV. SERVICE-ENTRANCE CONDUCTORS

230.40 Number of Service-Entrance Conductor Sets. Each service drop or lateral shall supply only one set of service-entrance conductors.

Exception No. 1: A building with one or more than one occupancy shall be permitted to have one set of service-entrance conductors for each service of different characteristics, as defined in 230.2(D), run to each occupancy or group of occupancies.

Exception No. 2: Where two to six service disconnecting means in separate enclosures are grouped at one location and supply separate loads from one service drop or lateral, one set of service-entrance conductors shall be permitted to supply each or several such service equipment enclosures.

Exception No. 3: A single-family dwelling unit and a separate structure shall be permitted to have one set of service-entrance conductors run to each from a single service drop or lateral.

Exception No. 4: A two-family dwelling or a multifamily dwelling shall be permitted to have one set of service-entrance conductors installed to supply the circuits covered in 210.25.

Exception No. 5: One set of service-entrance conductors connected to the supply side of the normal service disconnecting means shall be permitted to supply each or several systems covered by 230.82(4) or (5).

230.41 Insulation of Service-Entrance Conductors.
Service-entrance conductors entering or on the exterior of buildings or other structures shall be insulated.

Exception: A grounded conductor shall be permitted to be uninsulated as follows:

(a) *Bare copper used in a raceway or part of a service cable assembly.*
(b) *Bare copper for direct burial where bare copper is judged to be suitable for the soil conditions.*
(c) *Bare copper for direct burial without regard to soil conditions where part of a cable assembly identified for underground use.*
(d) *Aluminum or copper-clad aluminum without individual insulation or covering where part of a cable assembly or identified for underground use in a raceway, or for direct burial.*
(e) *Bare conductors used in an auxiliary gutter.*

230.42 Minimum Size and Rating.

(A) General. The ampacity of the service-entrance conductors before the application of any adjustment or correction factors shall not be less than either (1) or (2). Loads shall be determined in accordance with Article 220. Ampacity shall be determined from 310.15. The maximum allowable current of busways shall be that value for which the busway has been listed or labeled.

(1) The sum of the noncontinuous loads plus 125 percent of continuous loads

(2) The sum of the noncontinuous load plus the continuous load if the service-entrance conductors terminate in an overcurrent device where both the overcurrent device and its assembly are listed for operation at 100 percent of their rating

(B) Specific Installations. In addition to the requirements of 230.42(A), the minimum ampacity for ungrounded conductors for specific installations shall not be less than the rating of the service disconnecting means specified in 230.79(A) through (D).

(C) Grounded Conductors. The grounded conductor shall not be less than the minimum size as required by 250.24(B).

230.43 Wiring Methods for 600 Volts, Nominal, or Less. Service-entrance conductors shall be installed in accordance with the applicable requirements of this *Code* covering the type of wiring method used and shall be limited to the following methods:

(1)   Open wiring on insulators
(2)   Type IGS cable
(3)   Rigid metal conduit
(4)   Intermediate metal conduit
(5)   Electrical metallic tubing
(6)   Electrical nonmetallic tubing (ENT)
(7)   Service-entrance cables
(8)   Wireways
(9)   Busways
(10) Auxiliary gutters
(11) Rigid nonmetallic conduit
(12) Cablebus
(13) Type MC cable
(14) Mineral-insulated, metal-sheathed cable
(15) Flexible metal conduit not over 1.8 m (6 ft) long or liquidtight flexible metal conduit not over 1.8 m (6 ft) long between raceways, or between raceway and

service equipment, with equipment bonding jumper routed with the flexible metal conduit or the liquidtight flexible metal conduit according to the provisions of 250.102(A), (B), (C), and (E)

(16) Liquidtight flexible nonmetallic conduit

230.46 Spliced Conductors. Service-entrance conductors shall be permitted to be spliced or tapped in accordance with 110.14, 300.5(E), 300.13, and 300.15.

230.49 Protection Against Physical Damage—Underground. Underground service-entrance conductors shall be protected against physical damage in accordance with 300.5.

230.50 Protection of Open Conductors and Cables Against Damage—Above Ground. Service-entrance conductors installed above ground shall be protected against physical damage as specified in 230.50(A) or (B).

(A) Service Cables. Service cables, where subject to physical damage, shall be protected by any of the following:

(1) Rigid metal conduit
(2) Intermediate metal conduit
(3) Schedule 80 rigid nonmetallic conduit
(4) Electrical metallic tubing
(5) Other approved means

230.51 Mounting Supports. Cables or individual open service conductors shall be supported as specified in 230.51(A), (B), or (C).

(A) Service Cables. Service cables shall be supported by straps or other approved means within 300 mm (12 in.) of every service head, gooseneck, or connection to a raceway or enclosure and at intervals not exceeding 750 mm (30 in.).

230.53 Raceways to Drain. Where exposed to the weather, raceways enclosing service-entrance conductors shall be raintight and arranged to drain. Where embedded in masonry, raceways shall be arranged to drain.

230.54 Overhead Service Locations.

(A) Raintight Service Head. Service raceways shall be equipped with a raintight service head at the point of connection to service-drop conductors.

(B) Service Cable Equipped with Raintight Service Head or Gooseneck. Service cables shall be equipped with a raintight service head.

Exception: Type SE cable shall be permitted to be formed in a gooseneck and taped with a self-sealing weather-resistant thermoplastic.

(C) Service Heads Above Service-Drop Attachment. Service heads and goosenecks in service-entrance cables shall be located above the point of attachment of the service-drop conductors to the building or other structure.

Exception: Where it is impracticable to locate the service head above the point of attachment, the service head location shall be permitted not farther than 600 mm (24 in.) from the point of attachment.

(D) Secured. Service cables shall be held securely in place.

(E) Separately Bushed Openings. Service heads shall have conductors of different potential brought out through separately bushed openings.

Exception: For jacketed multiconductor service cable without splice.

(F) Drip Loops. Drip loops shall be formed on individual conductors. To prevent the entrance of moisture, service-entrance conductors shall be connected to the service-drop conductors either (1) below the level of the service head or (2) below the level of the termination of the service-entrance cable sheath.

(G) Arranged That Water Will Not Enter Service Raceway or Equipment. Service-drop conductors and service-entrance conductors shall be arranged so that water will not enter service raceway or equipment.

V. SERVICE EQUIPMENT—GENERAL

230.62 Service Equipment—Enclosed or Guarded. Energized parts of service equipment shall be enclosed as specified in 230.62(A) or guarded as specified in 230.62(B).

(A) Enclosed. Energized parts shall be enclosed so that they will not be exposed to accidental contact or shall be guarded as in 230.62(B).

230.66 Marking. Service equipment rated at 600 volts or less shall be marked to identify it as being suitable for use as service equipment. Individual meter socket enclosures shall not be considered service equipment.

VI. SERVICE EQUIPMENT—DISCONNECTING MEANS

230.70 General. Means shall be provided to disconnect all conductors in a building or other structure from the service-entrance conductors.

(A) Location. The service disconnecting means shall be installed in accordance with 230.70(A)(1), (2), and (3).

(1) Readily Accessible Location. The service disconnecting means shall be installed at a readily accessible location either outside of a building or structure or inside nearest the point of entrance of the service conductors.

(2) Bathrooms. Service disconnecting means shall not be installed in bathrooms.

(B) Marking. Each service disconnect shall be permanently marked to identify it as a service disconnect.

(C) Suitable for Use. Each service disconnecting means shall be suitable for the prevailing conditions. Service equipment installed in hazardous (classified) locations shall comply with the requirements of Articles 500 through 517.

230.71 Maximum Number of Disconnects.

(A) General. The service disconnecting means for each service permitted by 230.2, or for each set of service-entrance conductors permitted by 230.40, Exception Nos. 1, 3, 4, or 5, shall consist of not more than six switches or sets of circuit breakers, or a combination of not more than six switches and sets of circuit breakers, mounted in a single enclosure, in a group of separate enclosures, or in or on a switchboard. There shall be no more than six sets of disconnects per service grouped in any one location. For the purpose of this section, disconnecting means used solely for power monitoring equipment, or the control circuit of the ground-fault protection system or power-operable service disconnecting means, installed as part of the listed equipment, shall not be considered a service disconnecting means.

(B) Single-Pole Units. Two or three single-pole switches or breakers, capable of individual operation, shall be permitted on multiwire circuits, one pole for each ungrounded conductor, as one multipole disconnect, provided they are equipped with handle ties or a master handle to disconnect all conductors of the service with no more than six operations of the hand.

230.72 Grouping of Disconnects.

(A) General. The two to six disconnects as permitted in 230.71 shall be grouped. Each disconnect shall be marked to indicate the load served.

(C) Access to Occupants. In a multiple-occupancy building, each occupant shall have access to the occupant's service disconnecting means.

230.74 Simultaneous Opening of Poles. Each service disconnect shall simultaneously disconnect all ungrounded service conductors that it controls from the premises wiring system.

230.75 Disconnection of Grounded Conductor. Where the service disconnecting means does not disconnect the grounded conductor from the premises wiring, other means shall be provided for this purpose in the service equipment. A terminal or bus to which all grounded conductors can be attached by means of pressure connectors shall be permitted for this purpose. In a multisection switchboard, disconnects for the grounded conductor shall be permitted to be in any section of the switchboard, provided any such switchboard section is marked.

230.76 Manually or Power Operable. The service disconnecting means for ungrounded service conductors shall consist of either (1) a manually operable switch or circuit breaker equipped with a handle or other suitable operating means or (2) a power-operated switch or circuit breaker,

provided the switch or circuit breaker can be opened by hand in the event of a power supply failure.

230.77 Indicating. The service disconnecting means shall plainly indicate whether it is in the open or closed position.

230.79 Rating of Service Disconnecting Means. The service disconnecting means shall have a rating not less than the load to be carried, determined in accordance with Article 220. In no case shall the rating be lower than specified in 230.79(A), (B), (C), or (D).

(A) One-Circuit Installation. For installations to supply only limited loads of a single branch circuit, the service disconnecting means shall have a rating of not less than 15 amperes.

(B) Two-Circuit Installations. For installations consisting of not more than two 2-wire branch circuits, the service disconnecting means shall have a rating of not less than 30 amperes.

(C) One-Family Dwelling. For a one-family dwelling, the service disconnecting means shall have a rating of not less than 100 amperes, 3-wire.

(D) All Others. For all other installations, the service disconnecting means shall have a rating of not less than 60 amperes.

230.80 Combined Rating of Disconnects. Where the service disconnecting means consists of more than one switch or circuit breaker, as permitted by 230.71, the combined ratings of all the switches or circuit breakers used shall not be less than the rating required by 230.79.

230.81 Connection to Terminals. The service conductors shall be connected to the service disconnecting means by pressure connectors, clamps, or other approved means. Connections that depend on solder shall not be used.

230.82 Equipment Connected to the Supply Side of Service Disconnect. Only the following equipment shall be permitted to be connected to the supply side of the service disconnecting means:

(1) Cable limiters or other current-limiting devices.
(2) Meters, meter sockets, or meter disconnect switches nominally rated not in excess of 600 volts, provided all metal housings and service enclosures are grounded.

(3) Instrument transformers (current and voltage), high-impedance shunts, load management devices, and surge arresters.

(4) Taps used only to supply load management devices, circuits for standby power systems, fire pump equipment, and fire and sprinkler alarms, if provided with service equipment and installed in accordance with requirements for service-entrance conductors.

(5) Solar photovoltaic systems, fuel cell systems, or interconnected electric power production sources.

(6) Control circuits for power-operable service disconnecting means, if suitable overcurrent protection and disconnecting means are provided.

(7) Ground-fault protection systems where installed as part of listed equipment, if suitable overcurrent protection and disconnecting means are provided.

VII. SERVICE EQUIPMENT— OVERCURRENT PROTECTION

230.90 Where Required. Each ungrounded service conductor shall have overload protection.

(A) Ungrounded Conductor. Such protection shall be provided by an overcurrent device in series with each ungrounded service conductor that has a rating or setting not higher than the allowable ampacity of the conductor. A set of fuses shall be considered all the fuses required to protect all the ungrounded conductors of a circuit. Single-pole circuit breakers, grouped in accordance with 230.71(B), shall be considered as one protective device.

Exception No. 2: Fuses and circuit breakers with a rating or setting that conform with 240.4(B) or (C) and 240.6 shall be permitted.

Exception No. 3: Two to six circuit breakers or sets of fuses shall be permitted as the overcurrent device to provide the overload protection. The sum of the ratings of the circuit breakers or fuses shall be permitted to exceed the ampacity of the service conductors, provided the calculated load does not exceed the ampacity of the service conductors.

Exception No. 5: Overload protection for 120/240-volt, 3-wire, single-phase dwelling services shall be permitted in accordance with the requirements of 310.15(B)(6).

(B) Not in Grounded Conductor. No overcurrent device shall be inserted in a grounded service conductor except a circuit breaker that simultaneously opens all conductors of the circuit.

230.91 Location. The service overcurrent device shall be an integral part of the service disconnecting means or shall be located immediately adjacent thereto.

230.92 Locked Service Overcurrent Devices. Where the service overcurrent devices are locked or sealed or are not readily accessible to the occupant, branch-circuit overcurrent devices shall be installed on the load side, shall be mounted in a readily accessible location, and shall be of lower ampere rating than the service overcurrent device.

230.93 Protection of Specific Circuits. Where necessary to prevent tampering, an automatic overcurrent device that protects service conductors supplying only a specific load, such as a water heater, shall be permitted to be locked or sealed where located so as to be accessible.

230.94 Relative Location of Overcurrent Device and Other Service Equipment. The overcurrent device shall protect all circuits and devices.

CHAPTER 7

FEEDERS

INTRODUCTION

This chapter contains provisions from Article 215—Feeders. Article 215 covers installation, overcurrent protection, minimum size, and ampacity of conductors for feeders that supply branch-circuit loads. A feeder consists of all circuit conductors located between the service equipment, the source of a separately derived system, or other power supply source and the final branch-circuit overcurrent device. Feeder conductors supply power to remote panelboards, which are sometimes referred to as subpanels.

Although this article does not contain specifications for computing feeders, it does require compliance with Parts II, III, and IV of Article 220 (see Chapter 5 of this book).

ARTICLE 215
Feeders

215.1 Scope. This article covers the installation requirements, overcurrent protection requirements, minimum size, and ampacity of conductors for feeders supplying branch-circuit loads as computed in accordance with Article 220.

215.2 Minimum Rating and Size.

(A) Feeders Not More Than 600 Volts.

(1) General. Feeder conductors shall have an ampacity not less than required to supply the load as computed in Parts II, III, and IV of Article 220. The minimum feeder-circuit conductor size, before the application of any adjustment or correction factors, shall have an allowable ampacity not less than the noncontinuous load plus 125 percent of the continuous load.

Additional minimum sizes shall be as specified in (2), (3), and (4) under the conditions stipulated.

(2) For Specified Circuits. The ampacity of feeder conductors shall not be less than 30 amperes where the load supplied consists of any of the following number and types of circuits:

(1) Two or more 2-wire branch circuits supplied by a 2-wire feeder
(2) More than two 2-wire branch circuits supplied by a 3-wire feeder
(3) Two or more 3-wire branch circuits supplied by a 3-wire feeder

(3) Ampacity Relative to Service-Entrance Conductors. The feeder conductor ampacity shall not be less than that of the service-entrance conductors where the feeder conductors carry the total load supplied by service-entrance conductors with an ampacity of 55 amperes or less.

(4) Individual Dwelling Unit or Mobile Home Conductors. Feeder conductors for individual dwelling units or mobile homes need not be larger than service-entrance conductors. Paragraph 310.15(B)(6) shall be permitted to be used for conductor size.

215.3 Overcurrent Protection. Feeders shall be protected against overcurrent in accordance with the provisions of Part I of Article 240. Where a feeder supplies continuous loads or any combination of continuous and noncontinuous loads, the rating of the overcurrent device shall not be less than the noncontinuous load plus 125 percent of the continuous load.

215.4 Feeders with Common Neutral.

(A) Feeders with Common Neutral. Two or three sets of 3-wire feeders or two sets of 4-wire or 5-wire feeders shall be permitted to utilize a common neutral.

(B) In Metal Raceway or Enclosure. Where installed in a metal raceway or other metal enclosure, all conductors of all feeders using a common neutral shall be enclosed within the same raceway or other enclosure as required in 300.20.

215.5 Diagrams of Feeders. If required by the authority having jurisdiction, a diagram showing feeder details shall be provided prior to the installation of the feeders. Such a diagram shall show the area in square feet of the building or other structure supplied by each feeder, the total computed load before applying demand factors, the demand factors used, the computed load after applying demand factors, and the size and type of conductors to be used.

215.6 Feeder Conductor Grounding Means. Where a feeder supplies branch circuits in which equipment grounding conductors are required, the feeder shall include or provide a grounding means, in accordance with the provisions of 250.134, to which the equipment grounding conductors of the branch circuits shall be connected.

215.9 Ground-Fault Circuit-Interrupter Protection for Personnel. Feeders supplying 15- and 20-ampere receptacle branch circuits shall be permitted to be protected by a ground-fault circuit interrupter in lieu of the provisions for such interrupters as specified in 210.8 and Article 527.

CHAPTER 8

PANELBOARDS AND DISCONNECTS

INTRODUCTION

This chapter contains provisions from Article 408—Panelboards and Article 404—Switches. Article 408 covers panelboards installed for the control of light and power circuits. A panelboard consists of one or more panel units designed for assembly in the form of a single panel. It includes buses and automatic overcurrent devices and may contain switches for the control of light, heat, or power circuits. Panelboards are accessible only from the front and are designed to be placed in cabinets or cutout boxes. Panelboards can be used as service equipment (where identified for such use) or be installed as remote panelboards (subpanels).

Provisions from Article 404 are located in two chapters of this pocket guide, Chapters 8 and 18. This chapter covers switches and circuit breakers where used as switches. The switches discussed here are more commonly called disconnect switches or safety switches. Disconnects may contain overcurrent protective devices (fuses or circuit breakers), or they may be nonfusible or molded-case switches.

ARTICLE 408
Switchboards and Panelboards

I. GENERAL

408.1 Scope. This article covers the following:

(1) All switchboards, panelboards, and distribution boards installed for the control of light and power circuits
(2) Battery-charging panels supplied from light or power circuits

408.3 Support and Arrangement of Busbars and Conductors.

(C) Used as Service Equipment. Each switchboard or panelboard, if used as service equipment, shall be provided with a main bonding jumper sized in accordance with 250.28(D) or the equivalent placed within the panelboard or one of the sections of the switchboard for connecting the grounded service conductor on its supply side to the switchboard or panelboard frame.

408.4 Circuit Directory. All circuits and circuit modifications shall be legibly identified as to purpose or use on a circuit directory located on the face or inside of the panel door in the case of a panelboard, and at each switch on a switchboard.

III. PANELBOARDS

408.13 General. All panelboards shall have a rating not less than the minimum feeder capacity required for the load computed in accordance with Article 220. Panelboards shall be durably marked by the manufacturer with the voltage and the current rating and the number of phases for which they are designed and with the manufacturer's name or trademark in such a manner so as to be visible after installation, without disturbing the interior parts or wiring.

408.14 Classification of Panelboards. Panelboards shall be classified for the purposes of this article as either lighting and appliance branch-circuit panelboards or power panelboards, based on their content. A lighting and appliance branch circuit is a branch circuit that has a connection to the neutral of the panelboard and that has overcurrent protection of 30 amperes or less in one or more conductors.

(A) Lighting and Appliance Branch-Circuit Panelboard. A lighting and appliance branch-circuit panelboard is one having more than 10 percent of its overcurrent devices protecting lighting and appliance branch circuits.

(B) Power Panelboard. A power panelboard is one having 10 percent or fewer of its overcurrent devices protecting lighting and appliance branch circuits.

408.15 Number of Overcurrent Devices on One Panelboard. Not more than 42 overcurrent devices (other than those provided for in the mains) of a lighting and appliance branch-circuit panelboard shall be installed in any one cabinet or cutout box.

A lighting and appliance branch-circuit panelboard shall be provided with physical means to prevent the installation of more overcurrent devices than that number for which the panelboard was designed, rated, and approved.

For the purposes of this article, a 2-pole circuit breaker shall be considered two overcurrent devices; a 3-pole circuit breaker shall be considered three overcurrent devices.

408.16 Overcurrent Protection.

(A) Lighting and Appliance Branch-Circuit Panelboard Individually Protected. Each lighting and appliance branch-circuit panelboard shall be individually protected on the supply side by not more than two main circuit breakers or two sets of fuses having a combined rating not greater than that of the panelboard.

Exception No. 1: Individual protection for a lighting and appliance panelboard shall not be required if the panelboard feeder has overcurrent protection not greater than the rating of the panelboard.

Exception No. 2: For existing installations, individual protection for lighting and appliance branch-circuit panelboards shall not be required where such panelboards are used as service equipment in supplying an individual residential occupancy.

(B) Power Panelboard Protection. In addition to the requirements of 408.13, a power panelboard with supply conductors that include a neutral and having more than 10 percent of its overcurrent devices protecting branch circuits rated 30 amperes or less shall be protected by an overcurrent protective device having a rating not greater than that of the panelboard. The overcurrent protective device shall be located within or at any point on the supply side of the panelboard.

Exception: This individual protection shall not be required for a power panelboard used as service equipment with multiple disconnecting means in accordance with 230.71.

(F) Back-Fed Devices. Plug-in-type overcurrent protection devices or plug-in type-main lug assemblies that are back-fed and used to terminate field-installed ungrounded supply conductors shall be secured in place by an additional fastener that requires other than a pull to release the device from the mounting means on the panel.

408.17 Panelboards in Damp or Wet Locations. Panelboards in damp or wet locations shall be installed to comply with 312.2(A).

408.20 Grounding of Panelboards. Panelboard cabinets and panelboard frames, if of metal, shall be in physical contact with each other and shall be grounded. Where the panelboard is used with nonmetallic raceway or cable or where separate grounding conductors are provided, a terminal bar for the grounding conductors shall be secured inside the cabinet. The terminal bar shall be bonded to the cabinet and panelboard frame, if of metal; otherwise it shall be connected to the grounding conductor that is run with the conductors feeding the panelboard.

Grounding conductors shall not be connected to a terminal bar provided for grounded conductors (may be a neutral) unless the bar is identified for the purpose and is located where interconnection between equipment grounding conductors and grounded circuit conductors is permitted or required by Article 250.

408.21 Grounded Conductor Terminations. Each grounded conductor shall terminate within the panelboard in an individual terminal that is not also used for another conductor.

Exception: Grounded conductors of circuits with parallel conductors shall be permitted to terminate in a single terminal if the terminal is identified for connection of more than one conductor.

ARTICLE 404
Switches

404.3 Enclosure.

(A) General. Switches and circuit breakers shall be of the externally operable type mounted in an enclosure listed for the intended use. The minimum wire-bending space at terminals and minimum gutter space provided in switch enclosures shall be as required in 312.6.

(B) Used as a Raceway. Enclosures shall not be used as junction boxes, auxiliary gutters, or raceways for conductors feeding through or tapping off to other switches or overcurrent devices, unless the enclosure complies with 312.8.

404.4 Wet Locations. A switch or circuit breaker in a wet location or outside of a building shall be enclosed in a weatherproof enclosure or cabinet that shall comply with 312.2(A). Switches shall not be installed within wet locations in tub or shower spaces unless installed as part of a listed tub or shower assembly.

404.6 Position and Connection of Switches.

(A) Single-Throw Knife Switches. Single-throw knife switches shall be placed so that gravity will not tend to close them. Single-throw knife switches, approved for use in the inverted position, shall be provided with a locking device that ensures that the blades remain in the open position when so set.

404.7 Indicating. General-use and motor-circuit switches, circuit breakers, and molded case switches, where mounted in an enclosure as described in 404.3, shall clearly indicate whether they are in the open (off) or closed (on) position.

Where these switch or circuit breaker handles are operated vertically rather than rotationally or horizontally, the up position of the handle shall be the (on) position.

404.8 Accessibility and Grouping.

(A) Location. All switches and circuit breakers used as switches shall be located so that they may be operated from a readily accessible place. They shall be installed so that the center of the grip of the operating handle of the switch or circuit breaker, when in its highest position, is not more than 2.0 m (6 ft 7 in.) above the floor or working platform.

404.11 Circuit Breakers as Switches.
A hand-operable circuit breaker equipped with a lever or handle, or a power-operated circuit breaker capable of being opened by hand in the event of a power failure, shall be permitted to serve as a switch if it has the required number of poles.

CHAPTER 9

OVERCURRENT PROTECTION

INTRODUCTION

This chapter contains provisions from Article 240—Overcurrent Protection. Article 240 provides general requirements for overcurrent protection and overcurrent protective devices. Generally, overcurrent protective devices are plug fuses, cartridge fuses, and circuit breakers. Overcurrent protective devices are usually installed within enclosures, cabinets, or cutout boxes. Automatic overcurrent devices are part of the items that make up a panelboard.

Contained in this pocket guide are six of the nine parts that are included in Article 240. The six parts are: I General; II Location; III Enclosures; IV Disconnecting and Guarding; V Plug Fuses, Fuseholders, and Adapters; and VII Circuit Breakers.

ARTICLE 240
Overcurrent Protection

I. GENERAL

240.1 Scope. Parts I through VII of this article provide the general requirements for overcurrent protection and overcurrent protective devices not more than 600 volts, nominal.

240.4 Protection of Conductors. Conductors, other than flexible cords, flexible cables, and fixture wires, shall be protected against overcurrent in accordance with their ampacities specified in 310.15, unless otherwise permitted or required in 240.4(A) through (G).

(B) Devices Rated 800 Amperes or Less. The next higher standard overcurrent device rating (above the ampacity of the conductors being protected) shall be permitted to be used, provided all of the following conditions are met:

(1) The conductors being protected are not part of a multioutlet branch circuit supplying receptacles for cord-and-plug-connected portable loads.
(2) The ampacity of the conductors does not correspond with the standard ampere rating of a fuse or a circuit breaker without overload trip adjustments above its rating (but that shall be permitted to have other trip or rating adjustments).
(3) The next higher standard rating selected does not exceed 800 amperes.

(C) Devices Rated Over 800 Amperes. Where the overcurrent device is rated over 800 amperes, the ampacity of the conductors it protects shall be equal to or greater than the rating of the overcurrent device defined in 240.6.

(D) Small Conductors. Unless specifically permitted in 240.4(E) through (G), the overcurrent protection shall not exceed 15 amperes for 14 AWG, 20 amperes for 12 AWG, and 30 amperes for 10 AWG copper; or 15 amperes for 12 AWG and 25 amperes for 10 AWG aluminum and copper-clad aluminum after any correction factors for ambient temperature and number of conductors have been applied.

(E) Tap Conductors. Tap conductors shall be permitted to be protected against overcurrent in accordance with 210.19(A)(3) and (A)(4), 240.5(B)(2), 240.21, 368.11, 368.12, and 430.53(D).

240.6 Standard Ampere Ratings.

(A) Fuses and Fixed-Trip Circuit Breakers. The standard ampere ratings for fuses and inverse time circuit breakers shall be considered 15, 20, 25, 30, 35, 40, 45, 50, 60, 70, 80, 90, 100, 110, 125, 150, 175, 200, 225, 250, 300, 350, 400, 450, 500, 600, 700, 800, 1000, 1200, 1600, 2000, 2500, 3000, 4000, 5000, and 6000 amperes. Additional standard ampere ratings for fuses shall be 1, 3, 6, 10, and 601. The use of fuses and inverse time circuit breakers with nonstandard ampere ratings shall be permitted.

240.10 Supplementary Overcurrent Protection. Where supplementary overcurrent protection is used for luminaires (lighting fixtures), appliances, and other equipment or for internal circuits and components of equipment, it shall not be used as a substitute for branch-circuit overcurrent devices or in place of the branch-circuit protection specified in Article 210. Supplementary overcurrent devices shall not be required to be readily accessible.

II. LOCATION

240.20 Ungrounded Conductors.

(A) Overcurrent Device Required. A fuse or an overcurrent trip unit of a circuit breaker shall be connected in series with each ungrounded conductor.

(B) Circuit Breaker as Overcurrent Device. Circuit breakers shall open all ungrounded conductors of the circuit unless otherwise permitted in 240.20(B)(1), (B)(2), and (B)(3).

(1) Multiwire Branch Circuit. Except where limited by 210.4(B), individual single-pole circuit breakers, with or without approved handle ties, shall be permitted as the protection for each ungrounded conductor of multiwire branch circuits that serve only single-phase line-to-neutral loads.

(2) Grounded Single-Phase and 3-wire dc Circuits. In grounded systems, individual single-pole circuit breakers with approved handle ties shall be permitted as the protection for each ungrounded conductor for line-to-line connected loads for single-phase circuits or 3-wire, direct-current circuits.

240.21 Location in Circuit. Overcurrent protection shall be provided in each ungrounded circuit conductor and shall be located at the point where the conductors receive their supply except as specified in 240.21(A) through (G). No conductor supplied under the provisions of 240.21(A) through (G) shall supply another conductor under those provisions, except through an overcurrent protective device meeting the requirements of 240.4.

(A) Branch-Circuit Conductors. Branch-circuit tap conductors meeting the requirements specified in 210.19 shall be permitted to have overcurrent protection located as specified in that section.

(D) Service Conductors. Service-entrance conductors shall be permitted to be protected by overcurrent devices in accordance with 230.91.

240.23 Change in Size of Grounded Conductor. Where a change occurs in the size of the ungrounded conductor, a similar change shall be permitted to be made in the size of the grounded conductor.

240.24 Location in or on Premises.

(A) Accessibility. Overcurrent devices shall be readily accessible unless one of the following applies:

(2) For supplementary overcurrent protection, as described in 240.10.
(3) For overcurrent devices, as described in 225.40 and 230.92.
(4) For overcurrent devices adjacent to utilization equipment that they supply, access shall be permitted to be by portable means.

(B) Occupancy. Each occupant shall have ready access to all overcurrent devices protecting the conductors supplying that occupancy.

Exception No. 1: Where electric service and electrical maintenance are provided by the building management and where these are under continuous building management supervision, the service overcurrent devices and feeder overcurrent devices supplying more than one occupancy shall be permitted to be accessible to only authorized management personnel in the following:

(a) Multiple-occupancy buildings
(b) Guest rooms of hotels and motels that are intended for transient occupancy

Exception No. 2: Where electric service and electrical maintenance are provided by the building management and where these are under continuous building management supervision, the branch circuit overcurrent devices supplying any guest rooms shall be permitted to be accessible to only authorized management personnel for guest rooms of hotels and motels that are intended for transient occupancy.

(C) Not Exposed to Physical Damage. Overcurrent devices shall be located where they will not be exposed to physical damage.

(D) Not in Vicinity of Easily Ignitible Material. Overcurrent devices shall not be located in the vicinity of easily ignitible material, such as in clothes closets.

(E) Not Located in Bathrooms. In dwelling units and guest rooms of hotels and motels, overcurrent devices, other than supplementary overcurrent protection, shall not be located in bathrooms as defined in Article 100.

III. ENCLOSURES

240.32 Damp or Wet Locations. Enclosures for overcurrent devices in damp or wet locations shall comply with 312.2(A).

240.33 Vertical Position. Enclosures for overcurrent devices shall be mounted in a vertical position unless that is shown to be impracticable. Circuit breaker enclosures shall be permitted to be installed horizontally where the circuit breaker is installed in accordance with 240.81.

IV. DISCONNECTING AND GUARDING

240.40 Disconnecting Means for Fuses. A disconnecting means shall be provided on the supply side of all fuses in circuits over 150 volts to ground and cartridge fuses in circuits of any voltage where accessible to other than qualified persons so that each individual circuit containing fuses can be independently disconnected from the source of power. A current-limiting device without a disconnecting means shall be permitted on the supply side of the service disconnecting means as permitted by 230.82. A single disconnecting means shall be permitted on the supply side of more than one set of fuses as permitted by 430.112, Exception, for group operation of motors and 424.22(C) for fixed electric space-heating equipment.

V. PLUG FUSES, FUSEHOLDERS, AND ADAPTERS

240.50 General.

(A) Maximum Voltage. Plug fuses shall be permitted to be used in the following circuits:

(1) Circuits not exceeding 125 volts between conductors
(2) Circuits supplied by a system having a grounded neutral where the line-to-neutral voltage does not exceed 150 volts

(B) Marking. Each fuse, fuseholder, and adapter shall be marked with its ampere rating.

(C) Hexagonal Configuration. Plug fuses of 15-ampere and lower rating shall be identified by a hexagonal configuration of the window, cap, or other prominent part to distinguish them from fuses of higher ampere ratings.

(D) No Energized Parts. Plug fuses, fuseholders, and adapters shall have no exposed energized parts after fuses or fuses and adapters have been installed.

(E) Screw Shell. The screw shell of a plug-type fuseholder shall be connected to the load side of the circuit.

240.51 Edison-Base Fuses.

(A) Classification. Plug fuses of the Edison-base type shall be classified at not over 125 volts and 30 amperes and below.

(B) Replacement Only. Plug fuses of the Edison-base type shall be used only for replacements in existing installations where there is no evidence of overfusing or tampering.

240.52 Edison-Base Fuseholders. Fuseholders of the Edison-base type shall be installed only where they are made to accept Type S fuses by the use of adapters.

240.53 Type S Fuses. Type S fuses shall be of the plug type and shall comply with 240.53(A) and (B).

(A) Classification. Type S fuses shall be classified at not over 125 volts and 0 to 15 amperes, 16 to 20 amperes, and 21 to 30 amperes.

(B) Noninterchangeable. Type S fuses of an ampere classification as specified in 240.53(A) shall not be interchangeable with a lower ampere classification. They shall be designed so that they cannot be used in any fuseholder other than a Type S fuseholder or a fuseholder with a Type S adapter inserted.

240.54 Type S Fuses, Adapters, and Fuseholders.

(A) To Fit Edison-Base Fuseholders. Type S adapters shall fit Edison-base fuseholders.

(B) To Fit Type S Fuses Only. Type S fuseholders and adapters shall be designed so that either the fuseholder itself or the fuseholder with a Type S adapter inserted cannot be used for any fuse other than a Type S fuse.

(C) Nonremovable. Type S adapters shall be designed so that once inserted in a fuseholder, they cannot be removed.

(D) Nontamperable. Type S fuses, fuseholders, and adapters shall be designed so that tampering or shunting (bridging) would be difficult.

(E) Interchangeability. Dimensions of Type S fuses, fuseholders, and adapters shall be standardized to permit interchangeability regardless of the manufacturer.

VII. CIRCUIT BREAKERS

240.80 Method of Operation. Circuit breakers shall be trip free and capable of being closed and opened by manual operation. Their normal method of operation by other than manual means, such as electrical or pneumatic, shall be permitted if means for manual operation are also provided.

240.81 Indicating. Circuit breakers shall clearly indicate whether they are in the open "off" or closed "on" position.

Where circuit breaker handles are operated vertically rather than rotationally or horizontally, the "up" position of the handle shall be the "on" position.

240.83 Marking.

(A) Durable and Visible. Circuit breakers shall be marked with their ampere rating in a manner that will be durable and visible after installation. Such marking shall be permitted to be made visible by removal of a trim or cover.

(B) Location. Circuit breakers rated at 100 amperes or less and 600 volts or less shall have the ampere rating molded, stamped, etched, or similarly marked into their handles or escutcheon areas.

(C) Interrupting Rating. Every circuit breaker having an interrupting rating other than 5000 amperes shall have its interrupting rating shown on the circuit breaker. The interrupting rating shall not be required to be marked on circuit breakers used for supplementary protection.

CHAPTER 10

GROUNDED (NEUTRAL) CONDUCTORS

INTRODUCTION

This chapter contains provisions from Article 200—Use and Identification of Grounded Conductors. Article 200 covers grounded conductors in premises wiring systems and identification of grounded conductors and their terminals. A grounded conductor is a system or circuit conductor that is intentionally grounded. Grounded conductors are generally identified by a continuous white or gray outer finish or by three continuous white stripes on other than green insulation along its entire length. Although this is a short article, do not overlook the importance of proper use and identification of grounded conductors.

ARTICLE 200
Use and Identification of Grounded Conductors

200.1 Scope. This article provides requirements for the following:

(1) Identification of terminals
(2) Grounded conductors in premises wiring systems
(3) Identification of grounded conductors

200.2 General. All premises wiring systems, other than circuits and systems exempted or prohibited by 210.10, 215.7, 250.21, 250.22, 250.162, 503.13, 517.63, 668.11, 668.21, and 690.41, Exception, shall have a grounded conductor that is identified in accordance with 200.6.

The grounded conductor, where insulated, shall have insulation that is (1) suitable, other than color, for any ungrounded conductor of the same circuit on circuits of less than 1000 volts or impedance grounded neutral systems of 1 kV and over, or (2) rated not less than 600 volts for solidly grounded neutral systems of 1 kV and over as described in 250.184(A).

200.3 Connection to Grounded System. Premises wiring shall not be electrically connected to a supply system unless the latter contains, for any grounded conductor of the interior system, a corresponding conductor that is grounded. For the purpose of this section, *electrically connected* shall mean connected so as to be capable of carrying current, as distinguished from connection through electromagnetic induction.

200.6 Means of Identifying Grounded Conductors.

(A) Sizes 6 AWG or Smaller. An insulated grounded conductor of 6 AWG or smaller shall be identified by a continuous white or gray outer finish or by three continuous white stripes on other than green insulation along its entire length. Wires that have their outer covering finished to show a white or gray color but have colored tracer threads in the braid identifying the source of manufacture shall be considered as meeting the provisions of this section.

(B) Sizes Larger Than 6 AWG. An insulated grounded conductor larger than 6 AWG shall be identified either by a continuous white or gray outer finish or by three continuous white stripes on other than green insulation along its entire length or at the time of installation by a distinctive white marking at its terminations. This marking shall encircle the conductor or insulation.

(C) Flexible Cords. An insulated conductor that is intended for use as a grounded conductor, where contained within a flexible cord, shall be identified by a white or gray outer finish or by three continuous white stripes on other than green insulation or by methods permitted by 400.22.

(E) Grounded Conductors of Multiconductor Cables. The insulated grounded conductors in a multiconductor cable shall be identified by a continuous white or gray outer finish or by three continuous white stripes on other than green insulation along its entire length. Multiconductor flat 4 AWG or larger shall be permitted to employ an external ridge on the grounded conductor.

200.7 Use of Insulation of a White or Gray Color or with Three Continuous White Stripes.

(A) General. The following shall be used only for the grounded circuit conductor, unless otherwise permitted in 200.7(B) and (C):

(1) A conductor with continuous white or gray covering
(2) A conductor with three continuous white stripes on other than green insulation
(3) A marking of white or gray color at the termination

(B) Circuits of Less Than 50 Volts. A conductor with white or gray color insulation or three continuous white stripes or having a marking of white or gray at the termination for circuits of less than 50 volts shall be required to be grounded only as required by 250.20(A).

(C) Circuits of 50 Volts or More. The use of insulation that is white or gray or that has three continuous white stripes for other than a grounded conductor for circuits of

50 volts or more shall be permitted only as in (1) through (3).

(1) If part of a cable assembly and where the insulation is permanently reidentified to indicate its use as an ungrounded conductor, by painting or other effective means at its termination, and at each location where the conductor is visible and accessible.

(2) Where a cable assembly contains an insulated conductor for single-pole, 3-way or 4-way switch loops and the conductor with white or gray insulation or a marking of three continuous white stripes is used for the supply to the switch but not as a return conductor from the switch to the switched outlet. In these applications, the conductor with white or gray insulation or with three continuous white stripes shall be permanently reidentified to indicate its use by painting or other effective means at its terminations and at each location where the conductor is visible and accessible.

200.9 Means of Identification of Terminals. The identification of terminals to which a grounded conductor is to be connected shall be substantially white in color. The identification of other terminals shall be of a readily distinguishable different color.

CHAPTER 11

GROUNDING

INTRODUCTION

This chapter contains provisions from Article 250—Grounding. Article 250 covers general requirements for grounding and bonding of electrical installations: systems, circuits, equipment, and conductors. Having a good understanding of words and terms within Article 250 is imperative. Review the related terminology in Chapter 2 of this book, including bonding (bonded), bonding jumper, equipment bonding jumper, main bonding jumper, ground, grounded, effectively grounded, grounded conductor, grounding conductor, equipment grounding conductor, and grounding electrode conductor.

Contained in this pocket guide are seven of the ten parts that are included in Article 250. Those seven parts are: I General; II Circuit System Grounding; III Grounding Electrode System and Grounding Electrode Conductor; IV Enclosure, Raceway, and Service Cable Grounding; V Bonding; VI Equipment Grounding and Equipment Grounding Conductors; and VII Methods of Equipment Grounding.

ARTICLE 250
Grounding

I. GENERAL

250.1 Scope. This article covers general requirements for grounding and bonding of electrical installations, and specific requirements in (1) through (6).

(1) Systems, circuits, and equipment required, permitted, or not permitted to be grounded
(2) Circuit conductor to be grounded on grounded systems
(3) Location of grounding connections
(4) Types and sizes of grounding and bonding conductors and electrodes
(5) Methods of grounding and bonding
(6) Conditions under which guards, isolation, or insulation may be substituted for grounding.

250.4 General Requirements for Grounding and Bonding. The following general requirements identify what grounding and bonding of electrical systems are required to accomplish. The prescriptive methods contained in Article 250 shall be followed to comply with the performance requirements of this section.

(A) Grounded Systems.

(1) Electrical System Grounding. Electrical systems that are grounded shall be connected to earth in a manner that will limit the voltage imposed by lightning, line surges, or unintentional contact with higher-voltage lines and that will stabilize the voltage to earth during normal operation.

(2) Grounding of Electrical Equipment. Non–current-carrying conductive materials enclosing electrical conductors or equipment, or forming part of such equipment, shall be connected to earth so as to limit the voltage to ground on these materials.

(3) Bonding of Electrical Equipment. Non–current-carrying conductive materials enclosing electrical conductors or equipment, or forming part of such equipment, shall be connected together and to the electrical supply source in a manner that establishes an effective ground-fault current path.

(4) Bonding of Electrically Conductive Materials and Other Equipment. Electrically conductive materials that are likely to become energized shall be connected together and to the electrical supply source in a manner that establishes an effective ground-fault current path.

(5) Effective Ground-Fault Current Path. Electrical equipment and wiring and other electrically conductive material likely to become energized shall be installed in a manner that creates a permanent, low-impedance circuit capable of safely carrying the maximum ground-fault current likely to be imposed on it from any point on the wiring system where a ground fault may occur to the electrical supply source. The earth shall not be used as the sole equipment grounding conductor or effective ground-fault current path.

250.6 Objectionable Current over Grounding Conductors.

(A) Arrangement to Prevent Objectionable Current. The grounding of electrical systems, circuit conductors, surge arresters, and conductive non–current-carrying materials and equipment shall be installed and arranged in a manner that will prevent objectionable current over the grounding conductors or grounding paths.

(B) Alterations to Stop Objectionable Current. If the use of multiple grounding connections results in objectionable current, one or more of the following alterations shall be permitted to be made, provided that the requirements of 250.4(A)(5) or 250.4(B)(4) are met:

(1) Discontinue one or more but not all of such grounding connections.
(2) Change the locations of the grounding connections.
(3) Interrupt the continuity of the conductor or conductive path interconnecting the grounding connections.
(4) Take other suitable remedial and approved action.

250.8 Connection of Grounding and Bonding Equipment. Grounding conductors and bonding jumpers shall be connected by exothermic welding, listed pressure connectors, listed clamps, or other listed means. Connection devices or fittings that depend solely on solder shall not be used. Sheet metal screws shall not be used to connect grounding conductors to enclosures.

250.10 Protection of Ground Clamps and Fittings. Ground clamps or other fittings shall be approved for general use without protection or shall be protected from physical damage as indicated in (1) or (2).

(1) In installations where they are not likely to be damaged
(2) Where enclosed in metal, wood, or equivalent protective covering

250.12 Clean Surfaces. Nonconductive coatings (such as paint, lacquer, and enamel) on equipment to be grounded shall be removed from threads and other contact surfaces to ensure good electrical continuity or be connected by means of fittings designed so as to make such removal unnecessary.

II. CIRCUIT AND SYSTEM GROUNDING

250.20 Alternating-Current Circuits and Systems to Be Grounded. Alternating-current circuits and systems shall be grounded as provided for in 250.20(A), (B), (C), or (D). Other circuits and systems shall be permitted to be grounded. If such systems are grounded, they shall comply with the applicable provisions of this article.

(A) Alternating-Current Circuits of Less Than 50 Volts. Alternating-current circuits of less than 50 volts shall be grounded under any of the following conditions:

(1) Where supplied by transformers, if the transformer supply system exceeds 150 volts to ground
(2) Where supplied by transformers, if the transformer supply system is ungrounded
(3) Where installed as overhead conductors outside of buildings

(B) Alternating-Current Systems of 50 Volts to 1000 Volts. Alternating-current systems of 50 volts to 1000 volts that supply premises wiring and premises wiring systems shall be grounded under any of the following conditions:

(1) Where the system can be grounded so that the maximum voltage to ground on the ungrounded conductors does not exceed 150 volts

(2) Where the system is 3-phase, 4-wire, wye connected in which the neutral is used as a circuit conductor

(3) Where the system is 3-phase, 4-wire, delta connected in which the midpoint of one phase winding is used as a circuit conductor

250.24 Grounding Service-Supplied Alternating-Current Systems.

(A) System Grounding Connections. A premises wiring system supplied by a grounded ac service shall have a grounding electrode conductor connected to the grounded service conductor, at each service, in accordance with 250.24(A)(1) through (A)(5).

(1) General. The connection shall be made at any accessible point from the load end of the service drop or service lateral to and including the terminal or bus to which the grounded service conductor is connected at the service disconnecting means.

(4) Main Bonding Jumper as Wire or Busbar. Where the main bonding jumper specified in 250.28 is a wire or busbar and is installed from the neutral bar or bus to the equipment grounding terminal bar or bus in the service equipment, the grounding electrode conductor shall be permitted to be connected to the equipment grounding terminal bar or bus to which the main bonding jumper is connected.

(5) Load-Side Grounding Connections. A grounding connection shall not be made to any grounded circuit conductor on the load side of the service disconnecting means except as otherwise permitted in this article.

(B) Grounded Conductor Brought to Service Equipment. Where an ac system operating at less than 1000 volts is grounded at any point, the grounded conductor(s) shall be run to each service disconnecting means and shall be bonded to each disconnecting means enclosure. The grounded conductor(s) shall be installed in accordance with 250.24(B)(1) through (B)(3).

(1) Routing and Sizing. This conductor shall be routed with the phase conductors and shall not be smaller than the required grounding electrode conductor specified in Table 250.66 but shall not be required to be larger than the largest ungrounded service-entrance phase conductor. In addition, for service-entrance phase conductors larger than 1100 kcmil copper or 1750 kcmil aluminum, the grounded conductor shall not be smaller than 12-1/2 percent of the area of the largest service-entrance phase conductor.

(2) Parallel Conductors. Where the service-entrance phase conductors are installed in parallel, the size of the grounded conductor shall be based on the total circular mil area of the parallel conductors as indicated in this section. Where installed in two or more raceways, the size of the grounded conductor in each raceway shall be based on the size of the ungrounded service-entrance conductor in the raceway but not smaller than 1/0 AWG.

(C) Grounding Electrode Conductor. A grounding electrode conductor shall be used to connect the equipment grounding conductors, the service-equipment enclosures, and, where the system is grounded, the grounded service conductor to the grounding electrode(s) required by Part III of this article.

250.26 Conductor to Be Grounded—Alternating-Current Systems. For ac premises wiring systems, the conductor to be grounded shall be as specified in the following:

(1) Single-phase, 2-wire—one conductor
(2) Single-phase, 3-wire—the neutral conductor
(3) Multiphase systems having one wire common to all phases—the common conductor
(4) Multiphase systems where one phase is grounded—one phase conductor
(5) Multiphase systems in which one phase is used as in (2)—the neutral conductor

250.28 Main Bonding Jumper. For a grounded system, an unspliced main bonding jumper shall be used to connect the equipment grounding conductor(s) and the service-disconnect enclosure to the grounded conductor of the system within the enclosure for each service disconnect.

(A) Material. Main bonding jumpers shall be of copper or other corrosion-resistant material. A main bonding jumper shall be a wire, bus, screw, or similar suitable conductor.

(B) Construction. Where a main bonding jumper is a screw only, the screw shall be identified with a green finish that shall be visible with the screw installed.

(C) Attachment. Main bonding jumpers shall be attached in the manner specified by the applicable provisions of 250.8.

(D) Size. The main bonding jumper shall not be smaller than the sizes shown in Table 250.66 for grounding electrode conductors. Where the service-entrance phase conductors are larger than 1100 kcmil copper or 1750 kcmil aluminum, the bonding jumper shall have an area that is not less than 12-1/2 percent of the area of the largest phase conductor except that, where the phase conductors and the bonding jumper are of different materials (copper or aluminum), the minimum size of the bonding jumper shall be based on the assumed use of phase conductors of the same material as the bonding jumper and with an ampacity equivalent to that of the installed phase conductors.

250.32 Two or More Buildings or Structures Supplied from a Common Service.

See page 268 of Chapter 24.

III. GROUNDING ELECTRODE SYSTEM AND GROUNDING ELECTRODE CONDUCTOR

250.50 Grounding Electrode System. If available on the premises at each building or structure served, each item in 250.52(A)(1) through (A)(6) shall be bonded together to form the grounding electrode system. Where none of these electrodes are available, one or more of the electrodes specified in 250.52(A)(4) through (A)(7) shall be installed and used.

250.52 Grounding Electrodes.

(A) Electrodes Permitted for Grounding.

(1) Metal Underground Water Pipe. A metal underground water pipe in direct contact with the earth for 3.0 m (10 ft) or more (including any metal well casing effectively bonded to the pipe) and electrically continuous (or made electrically continuous by bonding around insulating joints or insulating pipe) to the points of connection of the grounding electrode conductor and the bonding conductors. Interior metal water piping located more than 1.52 m (5 ft) from the point of entrance to the building shall not be used as a part of the grounding electrode system or as a conductor to interconnect electrodes that are part of the grounding electrode system.

(2) Metal Frame of the Building or Structure. The metal frame of the building or structure, where effectively grounded.

(3) Concrete-Encased Electrode. An electrode encased by at least 50 mm (2 in.) of concrete, located within and near the bottom of a concrete foundation or footing that is in direct contact with the earth, consisting of at least 6.0 m (20 ft) of one or more bare or zinc galvanized or other electrically conductive coated steel reinforcing bars or rods of not less than 13 mm (1/2 in.) in diameter, or consisting of at least 6.0 m (20 ft) of bare copper conductor not smaller than 4 AWG. Reinforcing bars shall be permitted to be bonded together by the usual steel tie wires or other effective means.

(4) Ground Ring. A ground ring encircling the building or structure, in direct contact with the earth, consisting of at least 6.0 m (20 ft) of bare copper conductor not smaller than 2 AWG.

(5) Rod and Pipe Electrodes. Rod and pipe electrodes shall not be less than 2.5 m (8 ft) in length and shall consist of the following materials.

(a) Electrodes of pipe or conduit shall not be smaller than metric designator 21 (trade size 3/4) and, where of iron or steel, shall have the outer surface galvanized or otherwise metal-coated for corrosion protection.

(b) Electrodes of rods of iron or steel shall be at least 15.87 mm (5/8 in.) in diameter. Stainless steel rods less than 16 mm (5/8 in.) in diameter, nonferrous rods, or their equivalent shall be listed and shall not be less than 13 mm (1/2 in.) in diameter.

(6) Plate Electrodes. Each plate electrode shall expose not less than 0.186 m^2 (2 ft^2) of surface to exterior soil. Electrodes of iron or steel plates shall be at least 6.4 mm (1/4 in.) in thickness. Electrodes of nonferrous metal shall be at least 1.5 mm (0.06 in.) in thickness.

(7) Other Local Metal Underground Systems or Structures. Other local metal underground systems or structures such as piping systems and underground tanks.

(B) Electrodes Not Permitted for Grounding. The following shall not be used as grounding electrodes:

(1) Metal underground gas piping system
(2) Aluminum electrodes

250.53 Grounding Electrode System Installation.

(A) Rod, Pipe, and Plate Electrodes. Where practicable, rod, pipe, and plate electrodes shall be embedded below permanent moisture level. Rod, pipe, and plate electrodes shall be free from nonconductive coatings such as paint or enamel.

(B) Electrode Spacing. Where more than one of the electrodes of the type specified in 250.52(A)(5) or (A)(6) are used, each electrode of one grounding system (including that used for air terminals) shall not be less than 1.83 m (6 ft) from any other electrode of another grounding system. Two or more grounding electrodes that are effectively bonded together shall be considered a single grounding electrode system.

(C) Bonding Jumper. The bonding jumper(s) used to connect the grounding electrodes together to form the grounding electrode system shall be installed in accordance with 250.64(A), (B), and (E), shall be sized in accordance with 250.66, and shall be connected in the manner specified in 250.70.

(D) Metal Underground Water Pipe. Where used as a grounding electrode, metal underground water pipe shall meet the requirements of 250.53(D)(1) and (D)(2).

(1) Continuity. Continuity of the grounding path or the bonding connection to interior piping shall not rely on water meters or filtering devices and similar equipment.

(2) Supplemental Electrode Required. A metal underground water pipe shall be supplemented by an additional electrode of a type specified in 250.52(A)(2) through (A)(7). Where the supplemental electrode is a rod, pipe, or plate type, it shall comply with 250.56. The supplemental electrode shall be permitted to be bonded to the grounding electrode conductor, the grounded service-entrance conductor, the nonflexible grounded service raceway, or any grounded service enclosure.

Exception: The supplemental electrode shall be permitted to be bonded to the interior metal water piping at any convenient point as covered in 250.52(A)(1), Exception.

(E) Supplemental Electrode Bonding Connection Size. Where the supplemental electrode is a rod, pipe, or plate electrode, that portion of the bonding jumper that is the sole connection to the supplemental grounding electrode shall not be required to be larger than 6 AWG copper wire or 4 AWG aluminum wire.

(F) Ground Ring. The ground ring shall be buried at a depth below the earth's surface of not less than 750 mm (30 in.).

(G) Rod and Pipe Electrodes. The electrode shall be installed such that at least 2.44 m (8 ft) of length is in contact with the soil. It shall be driven to a depth of not less than 2.44 m (8 ft) except that, where rock bottom is encountered, the electrode shall be driven at an oblique angle not to exceed 45 degrees from the vertical or, where rock bottom is encountered at an angle up to 45 degrees, the electrode shall be permitted to be buried in a trench that is at least 750 mm (30 in.) deep. The upper end of the electrode shall be flush with or below ground level unless the aboveground end and the grounding electrode conductor attachment are protected against physical damage as specified in 250.10.

(H) Plate Electrode. Plate electrodes shall be installed not less than 750 mm (30 in.) below the surface of the earth.

250.54 Supplementary Grounding Electrodes. Supplementary grounding electrodes shall be permitted to be connected to the equipment grounding conductors specified in 250.118 and shall not be required to comply with the electrode bonding requirements of 250.50 or 250.53(C) or the resistance requirements of 250.56, but the earth shall not be used as the sole equipment grounding conductor.

250.56 Resistance of Rod, Pipe, and Plate Electrodes. A single electrode consisting of a rod, pipe, or plate that does not have a resistance to ground of 25 ohms or less shall be augmented by one additional electrode of any of the types specified by 250.52(A)(2) through (A)(7). Where multiple rod, pipe, or plate electrodes are installed to meet the requirements of this section, they shall not be less than 1.8 m (6 ft) apart.

FPN: The paralleling efficiency of rods longer than 2.5 m (8 ft) is improved by spacing greater than 1.8 m (6 ft).

250.58 Common Grounding Electrode. Where an ac system is connected to a grounding electrode in or at a building as specified in 250.24 and 250.32, the same electrode shall be used to ground conductor enclosures and equipment in or on that building. Where separate services supply a building and are required to be connected to a grounding electrode, the same grounding electrode shall be used.

Two or more grounding electrodes that are effectively bonded together shall be considered as a single grounding electrode system in this sense.

250.60 Use of Air Terminals. Air terminal conductors and driven pipes, rods, or plate electrodes used for grounding air terminals shall not be used in lieu of the grounding electrodes required by 250.50 for grounding wiring systems and equipment. This provision shall not prohibit the required bonding together of grounding electrodes of different systems.

250.62 Grounding Electrode Conductor Material. The grounding electrode conductor shall be of copper, aluminum, or copper-clad aluminum. The material selected shall be resistant to any corrosive condition existing at the installation or shall be suitably protected against corrosion. The conductor shall be solid or stranded, insulated, covered, or bare.

250.64 Grounding Electrode Conductor Installation. Grounding electrode conductors shall be installed as specified in 250.64(A) through (F).

(A) Aluminum or Copper-Clad Aluminum Conductors. Bare aluminum or copper-clad aluminum grounding conductors shall not be used where in direct contact with masonry or the earth or where subject to corrosive conditions. Where used outside, aluminum or copper-clad aluminum grounding conductors shall not be terminated within 450 mm (18 in.) of the earth.

(B) Securing and Protection from Physical Damage. A grounding electrode conductor or its enclosure shall be securely fastened to the surface on which it is carried. A 4 AWG copper or aluminum or larger conductor shall be protected if exposed to severe physical damage. A 6 AWG grounding conductor that is free from exposure to physical damage shall be permitted to be run along the surface of the building construction without metal covering or protection where it is securely fastened to the construction; otherwise, it shall be in rigid metal conduit, intermediate metal conduit, rigid nonmetallic conduit, electrical metallic tubing, or cable armor. Grounding conductors smaller than 6 AWG shall be in rigid metal conduit, intermediate metal conduit, rigid nonmetallic conduit, electrical metallic tubing, or cable armor.

(C) Continuous. The grounding electrode conductor shall be installed in one continuous length without a splice or joint, unless spliced only by irreversible compression-type connectors listed for the purpose or by the exothermic welding process.

(D) Grounding Electrode Conductor Taps. Where a service consists of more than a single enclosure as permitted in 230.40, Exception No. 2, it shall be permitted to connect taps to the grounding electrode conductor. Each such tap conductor shall extend to the inside of each such enclosure. The grounding electrode conductor shall be sized in accordance with 250.66, but the tap conductors shall be permitted to be sized in accordance with the grounding electrode conductors specified in 250.66 for the largest conductor serving the respective enclosures. The tap conductors shall be connected to the grounding electrode conductor in such a manner that the grounding electrode conductor remains without a splice.

(E) Enclosures for Grounding Electrode Conductors. Metal enclosures for grounding electrode conductors shall be electrically continuous from the point of attachment to cabinets or equipment to the grounding electrode and shall be securely fastened to the ground clamp or fitting. Metal enclosures that are not physically continuous from cabinet or equipment to the grounding electrode shall be made electrically continuous by bonding each end to the grounding electrode conductor. Where a raceway is used as protection for a grounding electrode conductor, the installation shall comply with the requirements of the appropriate raceway article.

250.66 Size of Alternating-Current Grounding Electrode Conductor. The size of the grounding electrode conductor of a grounded or ungrounded ac system shall not be less than given in Table 250.66, except as permitted in 250.66(A) through (C).

(A) Connections to Rod, Pipe, or Plate Electrodes. Where the grounding electrode conductor is connected to rod, pipe, or plate electrodes as permitted in 250.52(A)(5) or 250.52(A)(6), that portion of the conductor that is the sole connection to the grounding electrode shall not be

Table 250.66 Grounding Electrode Conductor for Alternating-Current Systems

Size of Largest Ungrounded Service-Entrance Conductor or Equivalent Area for Parallel Conductors[a] (AWG/kcmil)		Size of Grounding Electrode Conductor (AWG/kcmil)	
Copper	Aluminum or Copper-Clad Aluminum	Copper	Aluminum or Copper-Clad Aluminum[b]
2 or smaller	1/0 or smaller	8	6
1 or 1/0	2/0 or 3/0	6	4
2/0 or 3/0	4/0 or 250	4	2
Over 3/0 through 350	Over 250 through 500	2	1/0
Over 350 through 600	Over 500 through 900	1/0	3/0
Over 600 through 1100	Over 900 through 1750	2/0	4/0
Over 1100	Over 1750	3/0	250

Notes:

1. Where multiple sets of service-entrance conductors are used as permitted in 230.40, Exception No. 2, the equivalent size of the largest service-entrance conductor shall be determined by the largest sum of the areas of the corresponding conductors of each set.

2. Where there are no service-entrance conductors, the grounding electrode conductor size shall be determined by the equivalent size of the largest service-entrance conductor required for the load to be served.

[a]This table also applies to the derived conductors of separately derived ac systems

[b]See installation restrictions in 250.64(A).

required to be larger than 6 AWG copper wire or 4 AWG aluminum wire.

(B) Connections to Concrete-Encased Electrodes. Where the grounding electrode conductor is connected to a concrete-encased electrode as permitted in 250.52(A)(3), that portion of the conductor that is the sole connection to the grounding electrode shall not be required to be larger than 4 AWG copper wire.

(C) Connections to Ground Rings. Where the grounding electrode conductor is connected to a ground ring as permitted in 250.52(4), that portion of the conductor that is the sole connection to the grounding electrode shall not be required to be larger than the conductor used for the ground ring.

250.68 Grounding Electrode Conductor and Bonding Jumper Connection to Grounding Electrodes.

(A) Accessibility. The connection of a grounding electrode conductor or bonding jumper to a grounding electrode shall be accessible.

Exception: An encased or buried connection to a concrete-encased, driven, or buried grounding electrode shall not be required to be accessible.

(B) Effective Grounding Path. The connection of a grounding electrode conductor or bonding jumper to a grounding electrode shall be made in a manner that will ensure a permanent and effective grounding path. Where necessary to ensure the grounding path for a metal piping system used as a grounding electrode, effective bonding shall be provided around insulated joints and around any equipment likely to be disconnected for repairs or replacement. Bonding conductors shall be of sufficient length to permit removal of such equipment while retaining the integrity of the bond.

250.70 Methods of Grounding and Bonding Conductor Connection to Electrodes. The grounding or bonding conductor shall be connected to the grounding electrode by exothermic welding, listed lugs, listed pressure connectors, listed clamps, or other listed means. Connections depending on solder shall not be used. Ground clamps shall be listed for the materials of the grounding electrode and the grounding electrode conductor and, where used on pipe, rod, or other buried electrodes, shall also be listed for direct soil burial or concrete encasement. Not more than one conductor shall be connected to the grounding electrode by a single clamp or fitting unless the clamp or fitting is listed for multiple conductors. One of the following methods shall be used:

(1) A pipe fitting, pipe plug, or other approved device screwed into a pipe or pipe fitting
(2) A listed bolted clamp of cast bronze or brass, or plain or malleable iron
(3) For indoor telecommunications purposes only, a listed sheet metal strap-type ground clamp having a rigid metal base that seats on the electrode and having a strap of such material and dimensions that it is not likely to stretch during or after installation
(4) An equally substantial approved means

IV. ENCLOSURE, RACEWAY, AND SERVICE CABLE GROUNDING

250.80 Service Raceways and Enclosures. Metal enclosures and raceways for service conductors and equipment shall be grounded.

V. BONDING

250.90 General. Bonding shall be provided where necessary to ensure electrical continuity and the capacity to conduct safely any fault current likely to be imposed.

250.92 Services.

(A) Bonding of Services. The non–current-carrying metal parts of equipment indicated in 250.92(A)(1), (2), and (3) shall be effectively bonded together.

(1) The service raceways, cable trays, cablebus framework, auxiliary gutters, or service cable armor or sheath except as permitted in 250.84.
(2) All service enclosures containing service conductors, including meter fittings, boxes, or the like, interposed in the service raceway or armor.
(3) Any metallic raceway or armor enclosing a grounding electrode conductor as specified in 250.64(B). Bonding shall apply at each end and to all intervening race-

ways, boxes, and enclosures between the service equipment and the grounding electrode.

(B) Method of Bonding at the Service. Electrical continuity at service equipment, service raceways, and service conductor enclosures shall be ensured by one of the following methods:

(1) Bonding equipment to the grounded service conductor in a manner provided in 250.8
(2) Connections utilizing threaded couplings or threaded bosses on enclosures where made up wrenchtight
(3) Threadless couplings and connectors where made up tight for metal raceways and metal-clad cables
(4) Other approved devices, such as bonding-type locknuts and bushings

Bonding jumpers meeting the other requirements of this article shall be used around concentric or eccentric knockouts that are punched or otherwise formed so as to impair the electrical connection to ground. Standard locknuts or bushings shall not be the sole means for the bonding required by this section.

250.94 Bonding for Other Systems. An accessible means external to enclosures for connecting intersystem bonding and grounding conductors shall be provided at the service equipment and at the disconnecting means for any additional buildings or structures by at least one of the following means:

(1) Exposed nonflexible metallic raceways
(2) Exposed grounding electrode conductor
(3) Approved means for the external connection of a copper or other corrosion-resistant bonding or grounding conductor to the grounded raceway or equipment

250.96 Bonding Other Enclosures.

(A) General. Metal raceways, cable trays, cable armor, cable sheath, enclosures, frames, fittings, and other metal non–current-carrying parts that are to serve as grounding conductors, with or without the use of supplementary equipment grounding conductors, shall be effectively bonded where necessary to ensure electrical continuity and the capacity to conduct safely any fault current likely to be imposed on them. Any nonconductive paint, enamel, or similar coating shall be removed at threads, contact points, and contact surfaces or be connected by means of fittings designed so as to make such removal unnecessary.

250.102 Equipment Bonding Jumpers.

(A) Material. Equipment bonding jumpers shall be of copper or other corrosion-resistant material. A bonding jumper shall be a wire, bus, screw, or similar suitable conductor.

(B) Attachment. Equipment bonding jumpers shall be attached in the manner specified by the applicable provisions of 250.8 for circuits and equipment and by 250.70 for grounding electrodes.

(C) Size—Equipment Bonding Jumper on Supply Side of Service. The bonding jumper shall not be smaller than the sizes shown in Table 250.66 for grounding electrode conductors. Where the service-entrance phase conductors are larger than 1100 kcmil copper or 1750 kcmil aluminum, the bonding jumper shall have an area not less than 12-1/2 percent of the area of the largest phase conductor except that, where the phase conductors and the bonding jumper are of different materials (copper or aluminum), the minimum size of the bonding jumper shall be based on the assumed use of phase conductors of the same material as the bonding jumper and with an ampacity equivalent to that of the installed phase conductors. Where the service-entrance conductors are paralleled in two or more raceways or cables, the equipment bonding jumper, where routed with the raceways or cables, shall be run in parallel. The size of the bonding jumper for each raceway or cable shall be based on the size of the service-entrance conductors in each raceway or cable.

The bonding jumper for a grounding electrode conductor raceway or cable armor as covered in 250.64(E) shall be the same size or larger than the required enclosed grounding electrode conductor.

(D) Size—Equipment Bonding Jumper on Load Side of Service. The equipment bonding jumper on the load side of the service overcurrent devices shall be sized, as a minimum, in accordance with the sizes listed in Table 250.122, but shall not be required to be larger than the largest ungrounded circuit conductors supplying the equipment and shall not be smaller than 14 AWG.

A single common continuous equipment bonding jumper shall be permitted to bond two or more raceways or cables where the bonding jumper is sized in accordance with Table 250.122 for the largest overcurrent device supplying circuits therein.

(E) Installation. The equipment bonding jumper shall be permitted to be installed inside or outside of a raceway or enclosure. Where installed on the outside, the length of the equipment bonding jumper shall not exceed 1.8 m (6 ft) and shall be routed with the raceway or enclosure. Where installed inside of a raceway, the equipment bonding jumper shall comply with the requirements of 250.119 and 250.148.

Exception: An equipment bonding jumper longer than 1.8 m (6 ft) shall be permitted at outside pole locations for the purpose of bonding or grounding isolated sections of metal raceways or elbows installed in exposed risers of metal conduit or other metal raceway.

250.104 Bonding of Piping Systems and Exposed Structural Steel.

(A) Metal Water Piping. The metal water piping system shall be bonded as required in (1), (2), (3), or (4) of this section. The bonding jumper(s) shall be installed in accordance with 250.64(A), (B), and (E). The points of attachment of the bonding jumper(s) shall be accessible.

(1) General. Metal water piping system(s) installed in or attached to a building or structure shall be bonded to the service equipment enclosure, the grounded conductor at the service, the grounding electrode conductor where of sufficient size, or to the one or more grounding electrodes used. The bonding jumper(s) shall be sized in accordance with Table 250.66 except as permitted in 250.104(A)(2) and (A)(3).

(2) Buildings of Multiple Occupancy. In buildings of multiple occupancy where the metal water piping system(s) installed in or attached to a building or structure for the individual occupancies is metallically isolated from all other occupancies by use of nonmetallic water piping, the metal water piping system(s) for each occupancy shall be permitted to be bonded to the equipment grounding terminal of the panelboard or switchboard enclosure (other than service equipment) supplying that occupancy. The bonding jumper shall be sized in accordance with Table 250.122.

(3) Multiple Buildings or Structures Supplied from a Common Service. The metal water piping system(s) installed in or attached to a building or structure shall be bonded to the building or structure disconnecting means enclosure where located at the building or structure, to the equipment grounding conductor run with the supply conductors, or to the one or more grounding electrodes used. The bonding jumper(s) shall be sized in accordance with 250.66, based on the size of the feeder or branch circuit conductors that supply the building. The bonding jumper shall not be required to be larger than the largest ungrounded feeder or branch circuit conductor supplying the building.

(B) Other Metal Piping. Where installed in or attached to a building or structure, metal piping system(s), including gas piping, that may become energized shall be bonded to the service equipment enclosure, the grounded conductor at the service, the grounding electrode conductor where of sufficient size, or to the one or more grounding electrodes used. The bonding jumper(s) shall be sized in accordance with 250.122 using the rating of the circuit that may energize the piping system(s). The equipment grounding conductor for the circuit that may energize the piping shall be permitted to serve as the bonding jumper(s). The points of attachment of the bonding jumper(s) shall be accessible.

(C) Structural Steel. Exposed structural steel that is interconnected to form a steel building frame and is not intentionally grounded and may become energized shall be bonded to the service equipment enclosure, the grounded conductor at the service, the grounding electrode conductor where of sufficient size, or the one or more grounding electrodes used. The bonding jumper(s) shall be sized in accordance with Table 250.66 and installed in accordance with 250.64(A), (B), and (E). The points of attachment of the bonding jumper(s) shall be accessible.

VI. EQUIPMENT GROUNDING AND EQUIPMENT GROUNDING CONDUCTORS

250.110 Equipment Fastened in Place or Connected by Permanent Wiring Methods (Fixed). Exposed non–current-carrying metal parts of fixed equipment likely to become energized shall be grounded under any of the following conditions:

(1) Where within 2.5 m (8 ft) vertically or 1.5 m (5 ft) horizontally of ground or grounded metal objects and subject to contact by persons
(2) Where located in a wet or damp location and not isolated
(3) Where in electrical contact with metal

250.112 Fastened in Place or Connected by Permanent Wiring Methods (Fixed)—Specific. Exposed, non–current-carrying metal parts of the kinds of equipment described in 250.112(A) through (K), and non–current-carrying metal parts of equipment and enclosures described in 250.112(L) and (M), shall be grounded regardless of voltage.

(L) Motor-Operated Water Pumps. Motor-operated water pumps, including the submersible type.

(M) Metal Well Casings. Where a submersible pump is used in a metal well casing, the well casing shall be bonded to the pump circuit equipment grounding conductor.

250.114 Equipment Connected by Cord and Plug. Under any of the conditions described in (1) through (4), exposed non–current-carrying metal parts of cord-and-plug-connected equipment likely to become energized shall be grounded.

(3) In residential occupancies:

   a. Refrigerators, freezers, and air conditioners
   b. Clothes-washing, clothes-drying, dish-washing machines; kitchen waste disposers; information technology equipment; sump pumps and electrical aquarium equipment
   c. Hand-held motor-operated tools, stationary and fixed motor-operated tools, light industrial motor-operated tools

   d. Motor-operated appliances of the following types:
      hedge clippers, lawn mowers, snow blowers, and
      wet scrubbers
   e. Portable handlamps

250.118 Types of Equipment Grounding Conductors.
The equipment grounding conductor run with or enclosing
the circuit conductors shall be one or more or a combination of the following:

(1) A copper, aluminum, or copper-clad aluminum con-
     ductor. This conductor shall be solid or stranded; in-
     sulated, covered, or bare; and in the form of a wire or
     a busbar of any shape.
(2) Rigid metal conduit.
(3) Intermediate metal conduit.
(4) Electrical metallic tubing.
(5) Flexible metal conduit where both the conduit and fit-
     tings are listed for grounding.
(6) Listed flexible metal conduit that is not listed for
     grounding, meeting all the following conditions:

     a. The conduit is terminated in fittings listed for
        grounding.
     b. The circuit conductors contained in the conduit
        are protected by overcurrent devices rated at 20
        amperes or less.
     c. The combined length of flexible metal conduit and
        flexible metallic tubing and liquidtight flexible
        metal conduit in the same ground return path
        does not exceed 1.8 m (6 ft).
     d. The conduit is not installed for flexibility.

(7) Listed liquidtight flexible metal conduit meeting all the
     following conditions:

     a. The conduit is terminated in fittings listed for
        grounding.
     b. For metric designators 12 through 16 (trade sizes
        3/8 through 1/2), the circuit conductors contained
        in the conduit are protected by overcurrent de-
        vices rated at 20 amperes or less.
     c. For metric designators 21 through 35 (trade sizes
        3/4 through 1-1/4), the circuit conductors con-

tained in the conduit are protected by overcurrent devices rated not more than 60 amperes and there is no flexible metal conduit, flexible metallic tubing, or liquidtight flexible metal conduit in trade sizes metric designators 12 through 16 (trade sizes 3/8 through 1/2) in the grounding path.

d. The combined length of flexible metal conduit and flexible metallic tubing and liquidtight flexible metal conduit in the same ground return path does not exceed 1.8 m (6 ft).

e. The conduit is not installed for flexibility.

(8) Flexible metallic tubing where the tubing is terminated in fittings listed for grounding and meeting the following conditions:

a. The circuit conductors contained in the tubing are protected by overcurrent devices rated at 20 amperes or less.

b. The combined length of flexible metal conduit and flexible metallic tubing and liquidtight flexible metal conduit in the same ground return path does not exceed 1.8 m (6 ft).

(9) Armor of Type AC cable as provided in 320.108.

(10) The copper sheath of mineral-insulated, metal-sheathed cable.

(11) Type MC cable where listed and identified for grounding in accordance with the following:

a. The combined metallic sheath and grounding conductor of interlocked metal tape—type MC cable

b. The metallic sheath or the combined metallic sheath and grounding conductors of the smooth or corrugated tube type MC cable

250.119 Identification of Equipment Grounding Conductors. Unless required elsewhere in this *Code,* equipment grounding conductors shall be permitted to be bare, covered, or insulated. Individually covered or insulated equipment grounding conductors shall have a continuous outer finish that is either green or green with one or more yellow stripes except as permitted in this section.

(A) Conductors Larger Than 6 AWG. An insulated or covered conductor larger than 6 AWG copper or aluminum shall be permitted, at the time of installation, to be permanently identified as an equipment grounding conductor at each end and at every point where the conductor is accessible. Identification shall encircle the conductor and shall be accomplished by one of the following:

(1) Stripping the insulation or covering from the entire exposed length
(2) Coloring the exposed insulation or covering green
(3) Marking the exposed insulation or covering with green tape or green adhesive labels

250.120 Equipment Grounding Conductor Installation. An equipment grounding conductor shall be installed in accordance with 250.120(A), (B), and (C).

(A) Raceway, Cable Trays, Cable Armor, Cablebus, or Cable Sheaths. Where it consists of a raceway, cable tray, cable armor, cablebus framework, or cable sheath or where it is a wire within a raceway or cable, it shall be installed in accordance with the applicable provisions in this *Code* using fittings for joints and terminations approved for use with the type raceway or cable used. All connections, joints, and fittings shall be made tight using suitable tools.

(B) Aluminum and Copper-Clad Aluminum Conductors. Equipment grounding conductors of bare or insulated aluminum or copper-clad aluminum shall be permitted. Bare conductors shall not come in direct contact with masonry or the earth or where subject to corrosive conditions. Aluminum or copper-clad aluminum conductors shall not be terminated within 450 mm (18 in.) of the earth.

(C) Equipment Grounding Conductors Smaller Than 6 AWG. Equipment grounding conductors smaller than 6 AWG shall be protected from physical damage by a raceway or cable armor except where run in hollow spaces of walls or partitions, where not subject to physical damage, or where protected from physical damage.

250.122 Size of Equipment Grounding Conductors.

(A) General. Copper, aluminum, or copper-clad aluminum equipment grounding conductors of the wire type shall not be smaller than shown in Table 250.122 but shall not be required to be larger than the circuit conductors supplying the equipment. Where a raceway or a cable armor or sheath is used as the equipment grounding conductor, as provided in 250.118 and 250.134(A), it shall comply with 250.4(A)(5) or 250.4(B)(4).

(B) Increased in Size. Where ungrounded conductors are increased in size, equipment grounding conductors, where installed, shall be increased in size proportionately according to circular mil area of the ungrounded conductors.

(C) Multiple Circuits. Where a single equipment grounding conductor is run with multiple circuits in the same raceway or cable, it shall be sized for the largest overcurrent device protecting conductors in the raceway or cable.

(F) Conductors in Parallel. Where conductors are run in parallel in multiple raceways or cables as permitted in 310.4, the equipment grounding conductors, where used, shall be run in parallel in each raceway or cable. One of the methods in 250.122(F)(1) or (2) shall be used to ensure the equipment grounding conductors are protected.

(1) Each parallel equipment grounding conductor shall be sized on the basis of the ampere rating of the overcurrent device protecting the circuit conductors in the raceway or cable in accordance with Table 250.122.

250.126 Identification of Wiring Device Terminals. The terminal for the connection of the equipment grounding conductor shall be identified by one of the following:

(1) A green, not readily removable terminal screw with a hexagonal head.
(2) A green, hexagonal, not readily removable terminal nut.
(3) A green pressure wire connector. If the terminal for the grounding conductor is not visible, the conductor entrance hole shall be marked with the word *green* or *ground,* the letters *G* or *GR* or the grounding symbol shown in Figure 250.126, or otherwise identified by a distinctive green color. If the terminal for the equipment grounding conductor is readily removable, the area adjacent to the terminal shall be similarly marked.

Table 250.122 Minimum Size Equipment Grounding Conductors for Grounding Raceway and Equipment

Rating or Setting of Automatic Overcurrent Device in Circuit Ahead of Equipment, Conduit, etc., Not Exceeding (Amperes)	Size (AWG or kcmil)	
	Copper	Aluminum or Copper-Clad Aluminum*
15	14	12
20	12	10
30	10	8
40	10	8
60	10	8
100	8	6
200	6	4
300	4	2
400	3	1
500	2	1/0
600	1	2/0
800	1/0	3/0
1000	2/0	4/0
1200	3/0	250
1600	4/0	350
2000	250	400
2500	350	600
3000	400	600
4000	500	800
5000	700	1200
6000	800	1200

Note: Where necessary to comply with 250.4(A)(5) or 250.4(B)(4), the equipment grounding conductor shall be sized larger than given in this table.

*See installation restrictions in 250.120.

VII. METHODS OF EQUIPMENT GROUNDING

250.130 Equipment Grounding Conductor Connections. Equipment grounding conductor connections at the source of separately derived systems shall be made in accordance with 250.30(A)(1). Equipment grounding conductor connections at service equipment shall be made as indicated in 250.130(A) or (B). For replacement of non–grounding-type receptacles with grounding-type receptacles and for branch-circuit extensions only in existing installations that do not have an equipment grounding conductor in the branch circuit, connections shall be permitted as indicated in 250.130(C).

(A) For Grounded Systems. The connection shall be made by bonding the equipment grounding conductor to the grounded service conductor and the grounding electrode conductor.

(B) For Ungrounded Systems. The connection shall be made by bonding the equipment grounding conductor to the grounding electrode conductor.

(C) Nongrounding Receptacle Replacement or Branch Circuit Extensions. The equipment grounding conductor of a grounding-type receptacle or a branch-circuit extension shall be permitted to be connected to any of the following:

(1) Any accessible point on the grounding electrode system as described in 250.50
(2) Any accessible point on the grounding electrode conductor
(3) The equipment grounding terminal bar within the enclosure where the branch circuit for the receptacle or branch circuit originates
(4) For grounded systems, the grounded service conductor within the service equipment enclosure
(5) For ungrounded systems, the grounding terminal bar within the service equipment enclosure

250.132 Short Sections of Raceway. Isolated sections of metal raceway or cable armor, where required to be grounded, shall be grounded in accordance with 250.134.

250.134 Equipment Fastened in Place or Connected by Permanent Wiring Methods (Fixed)—Grounding. Unless grounded by connection to the grounded circuit conductor as permitted by 250.32, 250.140, and 250.142, non–current-carrying metal parts of equipment, raceways, and other enclosures, if grounded, shall be grounded by one of the following methods.

(A) Equipment Grounding Conductor Types. By any of the equipment grounding conductors permitted by 250.118.

(B) With Circuit Conductors. By an equipment grounding conductor contained within the same raceway, cable, or otherwise run with the circuit conductors.

250.138 Cord-and-Plug-Connected Equipment. Non–current-carrying metal parts of cord-and-plug-connected equipment, if grounded, shall be grounded by one of the methods in 250.138(A) or (B).

(A) By Means of an Equipment Grounding Conductor. By means of an equipment grounding conductor run with the power supply conductors in a cable assembly or flexible cord properly terminated in a grounding-type attachment plug with one fixed grounding contact.

(B) By Means of a Separate Flexible Wire or Strap. By means of a separate flexible wire or strap, insulated or bare, protected as well as practicable against physical damage, where part of equipment.

250.140 Frames of Ranges and Clothes Dryers. This section shall apply to existing branch-circuit installations only. New branch-circuit installations shall comply with 250.134 and 250.138. Frames of electric ranges, wall-mounted ovens, counter-mounted cooking units, clothes dryers, and outlet or junction boxes that are part of the circuit for these appliances shall be grounded in the manner specified by 250.134 or 250.138; or, except for mobile homes and recreational vehicles, shall be permitted to be grounded to the grounded circuit conductor if all the following conditions are met.

(1) The supply circuit is 120/240-volt, single-phase, 3-wire; or 208Y/120-volt derived from a 3-phase, 4-wire, wye-connected system.
(2) The grounded conductor is not smaller than 10 AWG copper or 8 AWG aluminum.
(3) The grounded conductor is insulated, or the grounded conductor is uninsulated and part of a Type SE service-entrance cable and the branch circuit originates at the service equipment.

(4) Grounding contacts of receptacles furnished as part of the equipment are bonded to the equipment.

250.142 Use of Grounded Circuit Conductor for Grounding Equipment.

(A) Supply-Side Equipment. A grounded circuit conductor shall be permitted to ground non–current-carrying metal parts of equipment, raceways, and other enclosures at any of the following locations:

(1) On the supply side or within the enclosure of the ac service-disconnecting means
(2) On the supply side or within the enclosure of the main disconnecting means for separate buildings as provided in 250.32(B)

(B) Load-Side Equipment. Except as permitted in 250.30(A)(1) and 250.32(B), a grounded circuit conductor shall not be used for grounding non–current-carrying metal parts of equipment on the load side of the service disconnecting means or on the load side of a separately derived system disconnecting means or the overcurrent devices for a separately derived system not having a main disconnecting means.

Exception No. 1: The frames of ranges, wall-mounted ovens, counter-mounted cooking units, and clothes dryers under the conditions permitted for existing installations by 250.140 shall be permitted to be grounded by a grounded circuit conductor.

Exception No. 2: It shall be permissible to ground meter enclosures by connection to the grounded circuit conductor on the load side of the service disconnect if

(a) No service ground-fault protection is installed, and
(b) All meter enclosures are located near the service disconnecting means, and
(c) The size of the grounded circuit conductor is not smaller than the size specified in Table 250.122 for equipment grounding conductors.

250.146 Connecting Receptacle Grounding Terminal to Box. An equipment bonding jumper shall be used to

connect the grounding terminal of a grounding-type receptacle to a grounded box unless grounded as in 250.146(A)
through (D).

(A) Surface Mounted Box. Where the box is mounted on
the surface, direct metal-to-metal contact between the device yoke and the box shall be permitted to ground the receptacle to the box. This provision shall not apply to cover-
mounted receptacles unless the box and cover combination are listed as providing satisfactory ground continuity
between the box and the receptacle.

(B) Contact Devices or Yokes. Contact devices or yokes
designed and listed for the purpose shall be permitted in
conjunction with the supporting screws to establish the
grounding circuit between the device yoke and flush-type
boxes.

(C) Floor Boxes. Floor boxes designed for and listed as
providing satisfactory ground continuity between the box
and the device shall be permitted.

(D) Isolated Receptacles. Where required for the reduction of electrical noise (electromagnetic interference) on
the grounding circuit, a receptacle in which the grounding
terminal is purposely insulated from the receptacle mounting means shall be permitted. The receptacle grounding
terminal shall be grounded by an insulated equipment
grounding conductor run with the circuit conductors. This
grounding conductor shall be permitted to pass through
one or more panelboards without connection to the panelboard grounding terminal as permitted in 408.20, Exception, so as to terminate within the same building or
structure directly at an equipment grounding conductor terminal of the applicable derived system or service.

> FPN: Use of an isolated equipment grounding con-
> ductor does not relieve the requirement for ground-
> ing the raceway system and outlet box.

**250.148 Continuity and Attachment of Equipment
Grounding Conductors to Boxes.** Where circuit conductors are spliced within a box, or terminated on equipment
within or supported by a box, any separate equipment

grounding conductors associated with those circuit conductors shall be spliced or joined within the box or to the box with devices suitable for the use. Connections depending solely on solder shall not be used. Splices shall be made in accordance with 110.14(B) except that insulation shall not be required. The arrangement of grounding connections shall be such that the disconnection or the removal of a receptacle, luminaire (fixture), or other device fed from the box will not interfere with or interrupt the grounding continuity.

Exception: The equipment grounding conductor permitted in 250.146(D) shall not be required to be connected to the other equipment grounding conductors or to the box.

(A) Metal Boxes. A connection shall be made between the one or more equipment grounding conductors and a metal box by means of a grounding screw that shall be used for no other purpose or a listed grounding device.

(B) Nonmetallic Boxes. One or more equipment grounding conductors brought into a nonmetallic outlet box shall be arranged so that a connection can be made to any fitting or device in that box requiring grounding.

CHAPTER 12

BOXES AND ENCLOSURES

INTRODUCTION

This chapter contains provisions from Article 314—Outlet, Device, Pull and Junction Boxes; Conduit Bodies; and Fittings. Article 314 covers the installation and use of all enclosures (and conduit bodies) used as outlet, device, pull, or junction boxes, depending on their use. Included also are installation requirements for fittings used to join raceways and to connect raceways (and cables) to boxes and conduit bodies. This article's provisions range from selecting the size and type of box to installing the cover or canopy.

Provisions for boxes enclosing 18 through 6 AWG conductors are located in 314.16. Boxes (outlet, device, and junction) covered by 314.16 are calculated from the sizes and numbers of conductors. Requirements for boxes containing larger conductors (4 AWG and larger) are in 314.28. Pull and junction boxes covered by 314.28 are calculated from the sizes and numbers of raceways.

ARTICLE 314

Outlet, Device, Pull, and Junction Boxes; Conduit Bodies; Fittings; and Manholes

I. SCOPE AND GENERAL

314.1 Scope. This article covers the installation and use of all boxes and conduit bodies used as outlet, device, junction, or pull boxes, depending on their use, and manholes and other electric enclosures intended for personnel entry. Cast, sheet metal, nonmetallic, and other boxes such as FS, FD, and larger boxes are not classified as conduit bodies. This article also includes installation requirements for fittings used to join raceways and to connect raceways and cables to boxes and conduit bodies.

314.2 Round Boxes. Round boxes shall not be used where conduits or connectors requiring the use of locknuts or bushings are to be connected to the side of the box.

314.3 Nonmetallic Boxes. Nonmetallic boxes shall be permitted only with open wiring on insulators, concealed knob-and-tube wiring, cabled wiring methods with entirely nonmetallic sheaths, flexible cords, and nonmetallic raceways.

Exception No. 1: Where internal bonding means are provided between all entries, nonmetallic boxes shall be permitted to be used with metal raceways or metal-armored cables.

Exception No. 2: Where integral bonding means with a provision for attaching an equipment bonding jumper inside the box are provided between all threaded entries in nonmetallic boxes listed for the purpose, nonmetallic boxes shall be permitted to be used with metal raceways or metal-armored cables.

314.4 Metal Boxes. All metal boxes shall be grounded in accordance with the provisions of Article 250.

314.5 Short-Radius Conduit Bodies. Conduit bodies such as capped elbows and service-entrance elbows that enclose conductors 6 AWG or smaller, and are only intended to enable the installation of the raceway and the contained conductors, shall not contain splices, taps, or devices and shall be of sufficient size to provide free space for all conductors enclosed in the conduit body.

II. INSTALLATION

314.15 Damp, Wet, or Hazardous (Classified) Locations.

(A) Damp or Wet Locations. In damp or wet locations, boxes, conduit bodies, and fittings shall be placed or equipped so as to prevent moisture from entering or accumulating within the box, conduit body, or fitting. Boxes, conduit bodies, and fittings installed in wet locations shall be listed for use in wet locations.

314.16 Number of Conductors in Outlet, Device, and Junction Boxes, and Conduit Bodies. Boxes and conduit bodies shall be of sufficient size to provide free space for all enclosed conductors. In no case shall the volume of the box, as calculated in 314.16(A), be less than the fill calculation as calculated in 314.16(B). The minimum volume for conduit bodies shall be as calculated in 314.16(C).

The provisions of this section shall not apply to terminal housings supplied with motors.

Boxes and conduit bodies enclosing conductors 4 AWG or larger shall also comply with the provisions of 314.28.

(A) Box Volume Calculations. The volume of a wiring enclosure (box) shall be the total volume of the assembled sections, and, where used, the space provided by plaster rings, domed covers, extension rings, and so forth, that are marked with their volume or are made from boxes the dimensions of which are listed in Table 314.16(A).

(1) Standard Boxes. The volumes of standard boxes that are not marked with their volume shall be as given in Table 314.16(A).

(2) Other Boxes. Boxes 1650 cm³ (100 in.³) or less, other than those described in Table 314.16(A), and nonmetallic boxes shall be durably and legibly marked by the manufacturer with their volume. Boxes described in Table 314.16(A)

Table 314.16(A) Metal Boxes

Box Trade Size			Minimum Volume		Maximum Number of Conductors*						
mm	in.		cm³	in.³	18	16	14	12	10	8	6
100 × 32	(4 × 1-1/4)	round/octagonal	205	12.5	8	7	6	5	5	5	2
100 × 38	(4 × 1-1/2)	round/octagonal	254	15.5	10	8	7	6	6	5	3
100 × 54	(4 × 2-1/8)	round/octagonal	353	21.5	14	12	10	9	8	7	4
100 × 32	(4 × 1-1/4)	square	295	18.0	12	10	9	8	7	6	3
100 × 38	(4 × 1-1/2)	square	344	21.0	14	12	10	9	8	7	4
100 × 54	(4 × 2-1/8)	square	497	30.3	20	17	15	13	12	10	6
120 × 32	(4-11/16 × 1-1/4)	square	418	25.5	17	14	12	11	10	8	5
120 × 38	(4-11/16 × 1-1/2)	square	484	29.5	19	16	14	13	11	9	5
120 × 54	(4-11/16 × 2-1/8)	square	689	42.0	28	24	21	18	16	14	8
75 × 50 × 38	(3 × 2 × 1-1/2)	device	123	7.5	5	4	3	3	3	2	1
75 × 50 × 50	(3 × 2 × 2)	device	164	10.0	6	5	5	4	4	3	2
75 × 50 × 57	(3 × 2 × 2-1/4)	device	172	10.5	7	6	5	4	4	3	2
75 × 50 × 65	(3 × 2 × 2-1/2)	device	205	12.5	8	7	6	5	5	4	2
75 × 50 × 70	(3 × 2 × 2-3/4)	device	230	14.0	9	8	7	6	5	4	2

Metric (mm)	Trade Size (in.)	Type	Min. cm³	Min. in.³	18	16	14	12	10	8	6
75 × 50 × 90	(3 × 2 × 3-1/2)	device	295	18.0	12	10	9	8	7	6	3
100 × 54 × 38	(4 × 2-1/8 × 1-1/2)	device	169	10.3	6	5	5	4	4	3	2
100 × 54 × 48	(4 × 2-1/8 × 1-7/8)	device	213	13.0	8	7	6	5	5	4	2
100 × 54 × 54	(4 × 2-1/8 × 2-1/8)	device	238	14.5	9	8	7	6	5	4	2
95 × 50 × 65	(3-3/4 × 2 × 2-1/2)	masonry box/gang	230	14.0	9	8	7	6	5	4	2
95 × 50 × 90	(3-3/4 × 2 × 3-1/2)	masonry box/gang	344	21.0	14	12	10	9	8	7	4
min. 44.5 depth	FS—single cover/gang (1-3/4)		221	13.5	9	7	6	6	5	4	2
min. 60.3 depth	FD—single cover/gang (2-3/8)		295	18.0	12	10	9	8	7	6	3
min. 44.5 depth	FS—multiple cover/gang (1-3/4)		295	18.0	12	10	9	8	7	6	3
min. 60.3 depth	FD—multiple cover/gang (2-3/8)		395	24.0	16	13	12	10	9	8	4

*Where no volume allowances are required by 314.16(B)(2) through 314.16(B)(5).

that have a volume larger than is designated in the table shall be permitted to have their volume marked as required by this section.

(B) Box Fill Calculations. The volumes in paragraphs 314.16(B)(1) through (5), as applicable, shall be added together. No allowance shall be required for small fittings such as locknuts and bushings.

(1) Conductor Fill. Each conductor that originates outside the box and terminates or is spliced within the box shall be counted once, and each conductor that passes through the box without splice or termination shall be counted once. The conductor fill shall be computed using Table 314.16(B). A conductor, no part of which leaves the box, shall not be counted.

Exception: An equipment grounding conductor or conductors or not over four luminaire (fixture) wires smaller than 14 AWG, or both, shall be permitted to be omitted from the calculations where they enter a box from a domed luminaire (fixture) or similar canopy and terminate within that box.

(2) Clamp Fill. Where one or more internal cable clamps, whether factory or field supplied, are present in the box, a single volume allowance in accordance with Table 314.16(B) shall be made based on the largest conductor present in the box. No allowance shall be required for a cable connector with its clamping mechanism outside the box.

(3) Support Fittings Fill. Where one or more luminaire (fixture) studs or hickeys are present in the box, a single volume allowance in accordance with Table 314.16(B) shall be made for each type of fitting based on the largest conductor present in the box.

(4) Device or Equipment Fill. For each yoke or strap containing one or more devices or equipment, a double volume allowance in accordance with Table 314.16(B) shall be made for each yoke or strap based on the largest conductor connected to a device(s) or equipment supported by that yoke or strap.

Table 314.16(B) Volume Allowance Required per Conductor

Size of Conductor (AWG)	Free Space Within Box for Each Conductor	
	cm^3	in.^{3}
18	24.6	1.50
16	28.7	1.75
14	32.8	2.00
12	36.9	2.25
10	41.0	2.50
8	49.2	3.00
6	81.9	5.00

(5) Equipment Grounding Conductor Fill. Where one or more equipment grounding conductors or equipment bonding jumpers enter a box, a single volume allowance in accordance with Table 314.16(B) shall be made based on the largest equipment grounding conductor or equipment bonding jumper present in the box. Where an additional set of equipment grounding conductors, as permitted by 250.146(D), is present in the box, an additional volume allowance shall be made based on the largest equipment grounding conductor in the additional set.

(C) Conduit Bodies.

(1) General. Conduit bodies enclosing 6 AWG conductors or smaller, other than short-radius conduit bodies as described in 314.5, shall have a cross-sectional area not less than twice the cross-sectional area of the largest conduit or tubing to which it is attached. The maximum number of conductors permitted shall be the maximum number permitted by Table 1 of Chapter 9 for the conduit or tubing to which it is attached.

(2) With Splices, Taps, or Devices. Only those conduit bodies that are durably and legibly marked by the manufacturer with their volume shall be permitted to contain splices, taps, or devices. The maximum number of conductors shall be computed in accordance with 314.16(B). Conduit bodies shall be supported in a rigid and secure manner.

314.17 Conductors Entering Boxes, Conduit Bodies, or Fittings. Conductors entering boxes, conduit bodies, or fittings shall be protected from abrasion and shall comply with 314.17(A) through (D).

(A) Openings to Be Closed. Openings through which conductors enter shall be adequately closed.

(B) Metal Boxes and Conduit Bodies. Where metal boxes or conduit bodies are installed with open wiring or concealed knob-and-tube wiring, conductors shall enter through insulating bushings or, in dry locations, through flexible tubing extending from the last insulating support to not less than 6 mm (1/4 in.) inside the box and beyond any cable clamps. Except as provided in 300.15(C), the wiring shall be firmly secured to the box or conduit body. Where raceway or cable is installed with metal boxes or conduit bodies, the raceway or cable shall be secured to such boxes and conduit bodies.

(C) Nonmetallic Boxes and Conduit Bodies. Nonmetallic boxes and conduit bodies shall be suitable for the lowest temperature-rated conductor entering the box. Where nonmetallic boxes and conduit bodies are used with open wiring or concealed knob-and-tube wiring, the conductors shall enter the box through individual holes. Where flexible tubing is used to enclose the conductors, the tubing shall extend from the last insulating support to not less than 6 mm (1/4 in.) inside the box and beyond any cable clamp. Where nonmetallic-sheathed cable or multiconductor Type UF cable is used, the sheath shall extend not less than 6 mm (1/4 in.) inside the box and beyond any cable clamp. In all instances, all permitted wiring methods shall be secured to the boxes.

Exception: Where nonmetallic-sheathed cable or multiconductor Type UF cable is used with single gang boxes not larger than a nominal size 57 mm × 100 mm (2-1/4 in. × 4 in.) mounted in walls or ceilings, and where the cable is fastened within 200 mm (8 in.) of the box measured along the sheath and where the sheath extends through a cable knockout not less than 6 mm (1/4 in.), securing the cable to the box shall not be required. Multiple cable entries shall be permitted in a single cable knockout opening.

(D) Conductors 4 AWG or Larger. Installation shall comply with 300.4(F).

314.19 Boxes Enclosing Flush Devices. Boxes used to enclose flush devices shall be of such design that the devices will be completely enclosed on back and sides and substantial support for the devices will be provided. Screws for supporting the box shall not be used in attachment of the device contained therein.

314.20 In Wall or Ceiling. In walls or ceilings with a surface of concrete, tile, gypsum, plaster, or other noncombustible material, boxes shall be installed so that the front edge of the box will not be set back of the finished surface more than 6 mm (1/4 in.).

In walls and ceilings constructed of wood or other combustible surface material, boxes shall be flush with the finished surface or project therefrom.

314.21 Repairing Plaster and Drywall or Plasterboard. Plaster, drywall, or plasterboard surfaces that are broken or incomplete shall be repaired so there will be no gaps or open spaces greater than 3 mm (1/8 in.) at the edge of the box or fitting.

314.22 Exposed Surface Extensions. Surface extensions from a flush-mounted box shall be made by mounting and mechanically securing an extension ring over the flush box. Equipment grounding and bonding shall be in accordance with Article 250.

Exception: A surface extension shall be permitted to be made from the cover of a flush-mounted box where the cover is designed so it is unlikely to fall off or be removed if its securing means becomes loose. The wiring method shall be flexible for a length sufficient to permit removal of the cover and provide access to the box interior, and arranged so that any bonding or grounding continuity is independent of the connection between the box and cover.

314.23 Supports. Enclosures within the scope of this article shall be supported in accordance with one or more of the provisions in 314.23(A) through (H).

(A) Surface Mounting. An enclosure mounted on a building or other surface shall be rigidly and securely fastened in place. If the surface does not provide rigid and secure support, additional support in accordance with other provisions of this section shall be provided.

(B) Structural Mounting. An enclosure supported from a structural member of a building or from grade shall be rigidly supported either directly or by using a metal, polymeric, or wood brace.

(1) Nails and Screws. Nails and screws, where used as a fastening means, shall be attached by using brackets on the outside of the enclosure, or they shall pass through the interior within 6 mm (1/4 in.) of the back or ends of the enclosure.

(2) Braces. Metal braces shall be protected against corrosion and formed from metal that is not less than 0.51 mm (0.020 in.) thick uncoated. Wood braces shall have a cross section not less than nominal 25 mm × 50 mm (1 in. × 2 in.). Wood braces in wet locations shall be treated for the conditions. Polymeric braces shall be identified as being suitable for the use.

(C) Mounting in Finished Surfaces. An enclosure mounted in a finished surface shall be rigidly secured thereto by clamps, anchors, or fittings identified for the application.

(D) Suspended Ceilings. An enclosure mounted to structural or supporting elements of a suspended ceiling shall be not more than 1650 cm^3 (100 in.^{3}) in size and shall be securely fastened in place in accordance with either (D)(1) or (D)(2).

(1) Framing Members. An enclosure shall be fastened to the framing members by mechanical means such as bolts, screws, or rivets, or by the use of clips or other securing means identified for use with the type of ceiling framing member(s) and enclosure(s) employed. The framing members shall be adequately supported and securely fastened to each other and to the building structure.

(2) Support Wires. The installation shall comply with the provisions of 300.11(A). The enclosure shall be secured, using methods identified for the purpose, to ceiling support wire(s), including any additional support wire(s) installed for that purpose. Support wire(s) used for enclosure support shall be fastened at each end so as to be taut within the ceiling cavity.

(E) Raceway Supported Enclosure, Without Devices, Luminaires (Fixtures), or Lampholders. An enclosure that does not contain a device(s) other than splicing devices or support a luminaire(s) [fixture(s)], lampholder, or other equipment and is supported by entering raceways shall not exceed 1650 cm^3 (100 in.^{3}) in size. It shall have threaded entries or have hubs identified for the purpose. It shall be supported by two or more conduits threaded wrenchtight into the enclosure or hubs. Each conduit shall be secured within 900 mm (3 ft) of the enclosure, or within 450 mm (18 in.) of the enclosure if all conduit entries are on the same side.

Exception: Rigid metal, intermediate metal, or rigid non-metallic conduit or electrical metallic tubing shall be permitted to support a conduit body of any size, including a conduit body constructed with only one conduit entry, provided the trade size of the conduit body is not larger than the largest trade size of the conduit or electrical metallic tubing.

(F) Raceway Supported Enclosures, with Devices, Luminaires (Fixtures), or Lampholders. An enclosure that contains a device(s) or supports a luminaire(s) [fixture(s)], lampholder, or other equipment and is supported by entering raceways shall not exceed 1650 cm^3 (100 in.^{3}) in size. It shall have threaded entries or have hubs identified for the purpose. It shall be supported by two or more conduits threaded wrenchtight into the enclosure or hubs. Each conduit shall be secured within 450 mm (18 in.) of the enclosure.

Exception No. 1: Rigid metal or intermediate metal conduit shall be permitted to support a conduit body of any size, including a conduit body constructed with only one conduit entry, provided the trade size of the conduit body is not larger than the largest trade size of the conduit.

(G) Enclosures in Concrete or Masonry. An enclosure supported by embedment shall be identified as suitably protected from corrosion and securely embedded in concrete or masonry.

(H) Pendant Boxes. An enclosure supported by a pendant shall comply with 314.23(H)(1) or (2).

(1) Flexible Cord. A box shall be supported from a multiconductor cord or cable in an approved manner that protects the conductors against strain, such as a strain-relief connector threaded into a box with a hub.

(2) Conduit. A box supporting lampholders or luminaires (lighting fixtures), or wiring enclosures within luminaires (fixtures) used in lieu of boxes in accordance with 300.15(B), shall be supported by rigid or intermediate metal conduit stems. For stems longer than 450 mm (18 in.), the stems shall be connected to the wiring system with flexible fittings suitable for the location. At the luminaire (fixture) end, the conduit(s) shall be threaded wrenchtight into the box or wiring enclosure, or into hubs identified for the purpose.

Where supported by only a single conduit, the threaded joints shall be prevented from loosening by the use of setscrews or other effective means, or the luminaire (fixture), at any point, shall be at least 2.5 m (8 ft) above grade or standing area and at least 900 mm (3 ft) measured horizontally to the 2.5 m (8 ft) elevation from windows, doors, porches, fire escapes, or similar locations. A luminaire (fixture) supported by a single conduit shall not exceed 300 mm (12 in.) in any horizontal direction from the point of conduit entry.

314.24 Depth of Outlet Boxes. No box shall have an internal depth of less than 12.7 mm (1/2 in.). Boxes intended to enclose flush devices shall have an internal depth of not less than 23.8 mm (15/16 in.).

314.25 Covers and Canopies. In completed installations, each box shall have a cover, faceplate, lampholder, or luminaire (fixture) canopy, except where the installation complies with 410.14(B).

(A) Nonmetallic or Metal Covers and Plates. Nonmetallic or metal covers and plates shall be permitted. Where metal covers or plates are used, they shall comply with the grounding requirements of 250.110.

(B) Exposed Combustible Wall or Ceiling Finish. Where a luminaire (fixture) canopy or pan is used, any combustible wall or ceiling finish exposed between the edge of the canopy or pan and the outlet box shall be covered with noncombustible material.

(C) Flexible Cord Pendants. Covers of outlet boxes and conduit bodies having holes through which flexible cord pendants pass shall be provided with bushings designed for the purpose or shall have smooth, well-rounded surfaces on which the cords may bear. So-called hard rubber or composition bushings shall not be used.

314.27 Outlet Boxes.

(A) Boxes at Luminaire (Lighting Fixture) Outlets. Boxes used at luminaire (lighting fixture) or lampholder outlets shall be designed for the purpose. At every outlet used exclusively for lighting, the box shall be designed or installed so that a luminaire (lighting fixture) may be attached.

Exception: A wall-mounted luminaire (fixture) weighing not more than 3 kg (6 lb) shall be permitted to be supported on other boxes or plaster rings that are secured to other boxes, provided the luminaire (fixture) or its supporting yoke is secured to the box with no fewer than two No. 6 or larger screws.

(B) Maximum Luminaire (Fixture) Weight. Outlet boxes or fittings installed as required by 314.23 shall be permitted to support luminaires (lighting fixtures) weighing 23 kg (50 lb) or less. A luminaire (lighting fixture) that weighs more than 23 kg (50 lb) shall be supported independently of the outlet box unless the outlet box is listed for the weight to be supported.

(C) Floor Boxes. Boxes listed specifically for this application shall be used for receptacles located in the floor.

(D) Boxes at Ceiling-Suspended (Paddle) Fan Outlets.
Where a box is used as the sole support of a ceiling-suspended (paddle) fan, the box shall be listed for the application and for the weight of the fan to be supported. The installation shall comply with 422.18.

314.28 Pull and Junction Boxes and Conduit Bodies.
Boxes and conduit bodies used as pull or junction boxes shall comply with 314.28(A) through (D).

(A) Minimum Size. For raceways containing conductors of 4 AWG or larger, and for cables containing conductors of 4 AWG or larger, the minimum dimensions of pull or junction boxes installed in a raceway or cable run shall comply with the following. Where an enclosure dimension is to be calculated based on the diameter of entering raceways, the diameter shall be the metric designator (trade size) expressed in the units of measurement employed.

(1) Straight Pulls. In straight pulls, the length of the box shall not be less than eight times the metric designator (trade size) of the largest raceway.

(2) Angle or U Pulls. Where splices or where angle or U pulls are made, the distance between each raceway entry inside the box and the opposite wall of the box shall not be less than six times the metric designator (trade size) of the largest raceway in a row. This distance shall be increased for additional entries by the amount of the sum of the diameters of all other raceway entries in the same row on the same wall of the box. Each row shall be calculated individually, and the single row that provides the maximum distance shall be used.

Exception: Where a raceway or cable entry is in the wall of a box or conduit body opposite a removable cover, the distance from that wall to the cover shall be permitted to comply with the distance required for one wire per terminal in Table 312.6(A).

The distance between raceway entries enclosing the same conductor shall not be less than six times the metric designator (trade size) of the larger raceway.

When transposing cable size into raceway size in 314.28(A)(1) and (A)(2), the minimum metric designator (trade size) raceway required for the number and size of conductors in the cable shall be used.

(3) Smaller Dimensions. Boxes or conduit bodies of dimensions less than those required in 314.28(A)(1) and (A)(2) shall be permitted for installations of combinations of conductors that are less than the maximum conduit or tubing fill (of conduits or tubing being used) permitted by Table 1 of Chapter 9, provided the box or conduit body has been listed for and is permanently marked with the maximum number and maximum size of conductors permitted.

314.29 Boxes and Conduit Bodies to Be Accessible. Boxes and conduit bodies shall be installed so that the wiring contained in them can be rendered accessible without removing any part of the building or, in underground circuits, without excavating sidewalks, paving, earth, or other substance that is to be used to establish the finished grade.

Exception: Listed boxes shall be permitted where covered by gravel, light aggregate, or noncohesive granulated soil if their location is effectively identified and accessible for excavation.

CHAPTER 13

CABLES

INTRODUCTION

This chapter contains provisions from Article 334—Nonmetallic-Sheathed Cable (Types NM, NMC, and NMS), Article 330—Metal Clad Cable (Type MC), Article 338—Service Entrance Cable (Types SE and USE), and Article 340—Underground Feeder and Branch-Circuit Cable (Type UF). Contained in each article are provisions pertaining to the use and installation for the cable listed. Covered in Part I of each article are the cable's scope and definition. Other topics covered in most of these articles include: uses (permitted and not permitted), bending radius, and ampacity.

While service-entrance cable used as service-entrance conductors must be installed as required by Article 230 where used for interior wiring it must comply with many of the installation requirements in Article 334. Underground feeder and branch-circuit cable installed as nonmetallic-sheathed cable must also comply with the provisions of Article 334.

Although the specifications within each article may be different, the arrangement and numbering in each article follow the same format. For example, the uses permitted for each cable are located in 3_ _.10, and the uses not permitted are located in 3_ _.12.

ARTICLE 334

Nonmetallic-Sheathed Cable: Types NM, NMC, and NMS

I. GENERAL

334.1 Scope. This article covers the use, installation, and construction specifications of nonmetallic-sheathed cable.

334.2 Definition.

Nonmetallic-Sheathed Cable. A factory assembly of two or more insulated conductors having an outer sheath of nonmetallic material.

II. INSTALLATION

334.10 Uses Permitted. Type NM, Type NMC, and Type NMS cables shall be permitted to be used in the following:

(1) One- and two-family dwellings.
(2) Multifamily dwellings permitted to be of Types III, IV, and V construction except as prohibited in 334.12.

(A) Type NM. Type NM cable shall be permitted as follows:

(1) For both exposed and concealed work in normally dry locations except as prohibited in 334.10(3).
(2) To be installed or fished in air voids in masonry block or tile walls

(B) Type NMC. Type NMC cable shall be permitted as follows:

(1) For both exposed and concealed work in dry, moist, damp, or corrosive locations, except as prohibited in 334.10(3)
(2) In outside and inside walls of masonry block or tile
(3) In a shallow chase in masonry, concrete, or adobe protected against nails or screws by a steel plate at least 1.59 mm (1/16 in.) thick and covered with plaster, adobe, or similar finish

(C) Type NMS. Type NMS cable shall be permitted as follows:

(1) For both exposed and concealed work in normally dry locations except as prohibited in 334.10(3)

(2) To be installed or fished in air voids in masonry block or tile walls

334.12 Uses Not Permitted.

(A) Types NM, NMC, and NMS. Types NM, NMC, and NMS cables shall not be used as follows:

(1) As open runs in dropped or suspended ceilings in other than one- and two-family and multifamily dwellings.

(2) As service-entrance cable.

(8) Embedded in poured cement, concrete, or aggregate.

(10) Types NM and NMS. Types NM and NMS cable shall not be used as follows:

   a. Where exposed to corrosive fumes or vapors
   b. Where embedded in masonry, concrete, adobe, fill, or plaster
   c. In a shallow chase in masonry, concrete, or adobe and covered with plaster, adobe, or similar finish
   d. Where exposed or subject to excessive moisture or dampness

334.15 Exposed Work. In exposed work, except as provided in 300.11(A), the cable shall be installed as specified in 334.15(A) through (C).

(A) To Follow Surface. The cable shall closely follow the surface of the building finish or of running boards.

(B) Protection from Physical Damage. The cable shall be protected from physical damage where necessary by conduit, electrical metallic tubing, Schedule 80 PVC rigid nonmetallic conduit, pipe, guard strips, listed surface metal or nonmetallic raceway, or other means. Where passing through a floor, the cable shall be enclosed in rigid metal conduit, intermediate metal conduit, electrical metallic tubing, Schedule 80 PVC rigid nonmetallic conduit, listed surface metal or nonmetallic raceway, or other metal pipe extending at least 150 mm (6 in.) above the floor.

(C) In Unfinished Basements. Where the cable is run at angles with joists in unfinished basements, it shall be permissible to secure cables not smaller than two 6 AWG or three 8 AWG conductors directly to the lower edges of the joists. Smaller cables shall be run either through bored holes in joists or on running boards.

334.17 Through or Parallel to Framing Members. Types NM, NMC, or NMS cable shall be protected in accordance with 300.4 where installed through or parallel to framing members. Grommets used as required in 300.4(B)(1) shall remain in place and be listed for the purpose of cable protection.

334.23 In Accessible Attics. The installation of cable in accessible attics or roof spaces shall also comply with 320.23.

320.23 In Accessible Attics. Type AC cables in accessible attics or roof spaces shall be installed as specified in 320.23(A) and (B).

(A) Where Run Across the Top of Floor Joists. Where run across the top of floor joists, or within 2.1 m (7 ft) of floor or floor joists across the face of rafters or studding, in attics and roof spaces that are accessible, the cable shall be protected by substantial guard strips that are at least as high as the cable. Where this space is not accessible by permanent stairs or ladders, protection shall only be required within 1.8 m (6 ft) of the nearest edge of the scuttle hole or attic entrance.

(B) Cable Installed Parallel to Framing Members. Where the cable is installed parallel to the sides of rafters, studs, or floor joists, neither guard strips nor running boards shall be required, and the installation shall also comply with 300.4(D).

334.24 Bending Radius. Bends in Types NM, NMC, and NMS cable shall be made so that the cable will not be damaged. The radius of the curve of the inner edge of any bend during or after installation shall not be less than five times the diameter of the cable.

334.30 Securing and Supporting. Nonmetallic-sheathed cable shall be secured by staples, cable ties, straps, hangers, or similar fittings designed and installed so as not to damage the cable at intervals not exceeding 1.4 m (4-1/2 ft) and within 300 mm (12 in.) of every cabinet, box, or fitting. Flat cables shall not be stapled on edge.

(A) Horizontal Runs through Holes and Notches. In other than vertical runs, cables installed in accordance with 300.4 shall be considered supported and secured where such support does not exceed 1.4-m (4-1/2-ft) intervals and the nonmetallic-sheathed cable is securely fastened in place by an approved means within 300 mm (12 in.) of each box, cabinet, conduit body, or other nonmetallic-sheathed cable termination.

(B) Unsupported Cables. Nonmetallic-sheathed cable shall be permitted to be unsupported where the cable:

(1) Is fished between access points, where concealed in finished buildings or finished panels for prefabricated buildings and supporting is impracticable
(2) Is not more than 1.4 m (4-1/2 ft) from the last point of support for connections within an accessible ceiling to luminaire(s) [lighting fixture(s)] or equipment

(C) Wiring Device Without a Separate Outlet Box. A wiring device identified for the use, without a separate outlet box, incorporating an integral cable clamp shall be permitted where the cable is secured in place at intervals not exceeding 1.4 m (4-1/2 ft) and within 300 mm (12 in.) from the wiring device wall opening, and there shall be at least a 300 mm (12 in.) loop of unbroken cable or 150 mm (6 in.) of a cable end available on the interior side of the finished wall to permit replacement.

334.40 Boxes and Fittings.

(A) Boxes of Insulating Material. Nonmetallic outlet boxes shall be permitted as provided in 314.3.

(B) Devices of Insulating Material. Switch, outlet, and tap devices of insulating material shall be permitted to be used without boxes in exposed cable wiring and for rewiring in existing buildings where the cable is concealed and fished. Openings in such devices shall form a close fit around the outer covering of the cable, and the device shall fully enclose the part of the cable from which any part of the covering has been removed. Where connections to conductors are by binding-screw terminals, there shall be available as many terminals as conductors.

(C) Devices with Integral Enclosures. Wiring devices with integral enclosures identified for such use shall be permitted as provided in 300.15(E).

334.80 Ampacity. The ampacity of Types NM, NMC, and NMS cable shall be determined in accordance with 310.15. The ampacity shall be in accordance with the 60°C (140°F) conductor temperature rating. The 90°C (194°F) rating shall be permitted to be used for ampacity derating purposes, provided the final derated ampacity does not exceed that for a 60°C (140°F) rated conductor. The ampacity of Types NM, NMC, and NMS cable installed in cable tray shall be determined in accordance with 392.11.

ARTICLE 330
Metal-Clad Cable: Type MC

I. GENERAL

330.1 Scope. This article covers the use, installation, and construction specifications of metal-clad cable, Type MC.

330.2 Definition.

Metal Clad Cable, Type MC. A factory assembly of one or more insulated circuit conductors with or without optical fiber members enclosed in an armor of interlocking metal tape, or a smooth or corrugated metallic sheath.

II. INSTALLATION

330.10 Uses Permitted.

(A) General Uses. Where not subject to physical damage, Type MC cables shall be permitted as follows:

(1) For services, feeders, and branch circuits
(2) For power, lighting, control, and signal circuits

(3) Indoors or outdoors

(4) Where exposed or concealed

(5) Direct buried where identified for such use

(7) In any raceway

(8) As open runs of cable

(9) As aerial cable on a messenger

(11) In dry locations and embedded in plaster finish on brick or other masonry except in damp or wet locations

(12) In wet locations where any of the following conditions are met:

    a. The metallic covering is impervious to moisture.

    b. A lead sheath or moisture-impervious jacket is provided under the metal covering.

    c. The insulated conductors under the metallic covering are listed for use in wet locations.

(B) Specific Uses. Type MC cable shall be installed in compliance with Articles 300, 490, 725, and 770.52 as applicable and in accordance with 330.10(B)(1) through (B)(4).

(2) Direct Buried. Direct-buried cable shall comply with 300.5 or 300.50, as appropriate.

(3) Installed as Service-Entrance Cable. Type MC cable installed as service-entrance cable shall comply with Article 230.

(4) Installed Outside of Buildings or as Aerial Cable. Type MC cable installed outside of buildings or as aerial cable shall comply with Article 225 and Article 396.

330.12 Uses Not Permitted. Type MC cable shall not be used where exposed to the following destructive corrosive conditions, unless the metallic sheath is suitable for the conditions or is protected by material suitable for the conditions:

(1) Direct burial in the earth

(2) In concrete

(3) Where exposed to cinder fills, strong chlorides, caustic alkalis, or vapors of chlorine or of hydrochloric acids

330.17 Through or Parallel to Framing Members. Type MC cable shall be protected in accordance with 300.4 where installed through or parallel to framing members.

330.23 In Accessible Attics. The installation of Type MC cable in accessible attics or roof spaces shall also comply with 320.23.

See 334.23.

330.24 Bending Radius. Bends in Type MC cable shall be made so that the cable will not be damaged. The radius of the curve of the inner edge of any bend shall not be less than shown in 330.24(A) through (C).

(A) Smooth Sheath.

(1) Ten times the external diameter of the metallic sheath for cable not more than 19 mm (3/4 in.) in external diameter

(2) Twelve times the external diameter of the metallic sheath for cable more than 19 mm (3/4 in.) but not more than 38 mm (1-1/2 in.) in external diameter

(3) Fifteen times the external diameter of the metallic sheath for cable more than 38 mm (1-1/2 in.) in external diameter

(B) Interlocked-Type Armor or Corrugated Sheath. Seven times the external diameter of the metallic sheath.

(C) Shielded Conductors. Twelve times the overall diameter of one of the individual conductors or seven times the overall diameter of the multiconductor cable, whichever is greater.

330.30 Securing and Supporting. Type MC cable shall be supported and secured at intervals not exceeding 1.8 m (6 ft).

(A) Horizontal Runs Through Holes and Notches. In other than vertical runs, cables installed in accordance with 300.4 shall be considered supported and secured where such support does not exceed 1.8-m (6-ft) intervals.

(B) Unsupported Cables. Type MC cable shall be permitted to be unsupported where the cable:

(1) Is fished between access points, where concealed in finished buildings or structures and supporting is impractical

(2) Is not more than 1.8 m (6 ft) from the last point of support for connections within an accessible ceiling to luminaire(s) [lighting fixture(s)] or equipment

(C) At Terminations. Cables containing four or fewer conductors, sized no larger than 10 AWG, shall be secured within 300 mm (12 in.) of every box, cabinet, fitting, or other cable termination.

330.40 Boxes and Fitting. Fittings used for connecting Type MC cable to boxes, cabinets, or other equipment shall be listed and identified for such use.

330.80 Ampacity. The ampacity of Type MC cable shall be determined in accordance with 310.15 or 310.60 for 14 AWG and larger conductors and in accordance with Table 402.5 for 18 AWG and 16 AWG conductors. The installation shall not exceed the temperature ratings of terminations and equipment.

ARTICLE 338
Service-Entrance Cable: Types SE and USE

I. GENERAL

338.1 Scope. This article covers the use, installation, and construction specifications of service-entrance cable.

338.2 Definitions.

Service-Entrance Cable. A single conductor or multiconductor assembly provided with or without an overall covering, primarily used for services, and of the following types:

Type SE. Service-entrance cable having a flame-retardant, moisture-resistant covering.

Type USE. Service-entrance cable, identified for underground use, having a moisture-resistant covering, but not required to have a flame-retardant covering.

II. INSTALLATION

338.10 Uses Permitted.

(A) Service-Entrance Conductors. Service-entrance cable used as service-entrance conductors shall be installed as required by Article 230.

Type USE used for service laterals shall be permitted to emerge from the ground outside at terminations in meter bases or other enclosures where protected in accordance with 300.5(D).

(B) Branch Circuits or Feeders.

(1) Grounded Conductor Insulated. Type SE service-entrance cables shall be permitted in wiring systems where all of the circuit conductors of the cable are of the rubber-covered or thermoplastic type.

(2) Grounded Conductor Not Insulated. Type SE service-entrance cable shall be permitted for use where the insulated conductors are used for circuit wiring and the uninsulated conductor is used only for equipment grounding purposes.

Exception: Uninsulated conductors shall be permitted as a grounded conductor in accordance with 250.140.

(3) Temperature Limitations. Type SE service-entrance cable used to supply appliances shall not be subject to conductor temperatures in excess of the temperature specified for the type of insulation involved.

(4) Installation Methods for Branch Circuits and Feeders.

(a) Interior Installations. In addition to the provisions of this article, Type SE service-entrance cable used for interior wiring shall comply with the installation requirements of Parts I and II of Article 334, excluding 334.80.

(b) Exterior Installations. In addition to the provisions of this article, service-entrance cable used for feeders or branch circuits, where installed as exterior wiring, shall be installed as required by Article 225. The cable shall be supported in accordance with 334.30, unless used as messenger-supported wiring as allowed by Article 396.

Type USE cable shall be installed outside in accordance with the provisions of Article 340. Type USE shall be permitted to be terminated in enclosures at an indoor location where Type USE cable emerges from the ground. The length of the cable extending indoors to the first termination box shall not exceed 1.8 m (6 ft). Where Type USE cable emerges from the ground at terminations, it shall be protected in accordance with 300.5(D). Multiconductor service-entrance cable shall be permitted to be installed as messenger-supported wiring in accordance with Articles 225 and 396.

338.24 Bending Radius. Bends in Types USE and SE cable shall be made so that the cable will not be damaged. The radius of the curve of the inner edge of any bend, during or after installation, shall not be less than five times the diameter of the cable.

ARTICLE 340

Underground Feeder and Branch-Circuit Cable: Type UF

I. GENERAL

340.1 Scope. This article covers the use, installation, and construction specifications for underground feeder and branch-circuit cable, Type UF.

340.2 Definition.

Underground Feeder and Branch-Circuit Cable, Type UF. A listed factory assembly of one or more insulated conductors with an integral or an overall covering of nonmetallic material suitable for direct burial in the earth.

II. INSTALLATION

340.10 Uses Permitted. Type UF cable shall be permitted as follows:

(1) For use underground, including direct burial in the earth. For underground requirements, see 300.5.
(2) As single-conductor cables. Where installed as single-conductor cables, all conductors of the feeder grounded

conductor or branch circuit, including the grounded conductor and equipment grounding conductor, if any, shall be installed in accordance with 300.3.

(3) For wiring in wet, dry, or corrosive locations under the recognized wiring methods of this *Code*.

(4) Installed as nonmetallic-sheathed cable. Where so installed, the installation and conductor requirements shall comply with the provisions of Article 334 and shall be of the multiconductor type.

(5) For solar photovoltaic systems in accordance with 690.31.

(6) As single-conductor cables as the nonheating leads for heating cables as provided in 424.43.

(7) Supported by cable trays. Type UF cable supported by cable trays shall be of the multiconductor type.

340.12 Uses Not Permitted. Type UF cable shall not be used as follows:

(1) As service-entrance cable

(8) Embedded in poured cement, concrete, or aggregate, except where embedded in plaster as nonheating leads where permitted in 424.43

(9) Where exposed to direct rays of the sun, unless identified as sunlight resistant

(10) Where subject to physical damage

(11) As overhead cable, except where installed as messenger-supported wiring in accordance with Article 396

340.24 Bending Radius. Bends in Type UF cable shall be made so that the cable shall not be damaged. The radius of the curve of the inner edge of any bend shall not be less than five times the diameter of the cable.

340.80 Ampacity. The ampacity of Type UF cable shall be that of 60°C (140°F) conductors in accordance with 310.15.

CHAPTER 14

RACEWAYS

INTRODUCTION

This chapter contains provisions from Article 358—Electrical Metallic Tubing (Type EMT), Article 344—Rigid Metal Conduit (Type RMC), Article 352—Rigid Nonmetallic Conduit (Type RNC), Article 362—Electrical Nonmetallic Tubing (Type ENT), Article 348—Flexible Metal Conduit (Type FMC), Article 350—Liquidtight Flexible Metal Conduit (Type LFMC), and Article 356—Liquidtight Flexible Nonmetallic Conduit (Type LFNC). Contained in each article are provisions pertaining to the use and installation for the conduit or tubing listed. Part I of each article covers the raceway's scope and definition. Other topics covered in most of these articles include: uses (permitted and not permitted), size (minimum and maximum), number of conductors, bends (how made and number in one run), securing and supporting, and grounding.

Although the requirements in each article may be different, the arrangement and numbering in each article follow the same format. For example, in each article, 3_ _.10 covers uses permitted, 3_ _.12 covers uses not permitted, 3_ _.20 covers size, 3_ _.22 covers number of conductors, 3_ _.30 covers securing and supporting, and 3_ _.60 covers grounding.

ARTICLE 358
Electrical Metallic Tubing: Type EMT

I. GENERAL

358.1 Scope. This article covers the use, installation, and construction specifications for electrical metallic tubing (EMT) and associated fittings.

358.2 Definition.

Electrical Metallic Tubing (EMT). An unthreaded thin-wall raceway of circular cross section designed for the physical protection and routing of conductors and cables and for use as an equipment grounding conductor when installed utilizing appropriate fittings. EMT is generally made of steel (ferrous) with protective coatings or aluminum (nonferrous).

II. INSTALLATION

358.10 Uses Permitted.

(A) Exposed and Concealed. The use of EMT shall be permitted for both exposed and concealed work.

(B) Corrosion Protection. Ferrous or nonferrous EMT, elbows, couplings, and fittings shall be permitted to be installed in concrete, in direct contact with the earth, or in areas subject to severe corrosive influences where protected by corrosion protection and judged suitable for the condition.

(C) Wet Locations. All supports, bolts, straps, screws, and so forth shall be of corrosion-resistant materials or protected against corrosion by corrosion-resistant materials.

358.12 Uses Not Permitted. EMT shall not be used under the following conditions:

(1) Where, during installation or afterward, it will be subject to severe physical damage
(2) Where protected from corrosion solely by enamel
(3) In cinder concrete or cinder fill where subject to permanent moisture unless protected on all sides by a layer of noncinder concrete at least 50 mm (2 in.) thick or

unless the tubing is at least 450 mm (18 in.) under the fill

(5) For the support of luminaires (fixtures) or other equipment except conduit bodies no larger than the largest trade size of the tubing

358.24 Bends—How Made. Bends shall be made so that the tubing is not damaged and the internal diameter of the tubing is not effectively reduced. The radius of the curve of any field bend to the centerline of the conduit shall not be less than shown in Table 344.24 for one-shot and full shoe benders.

358.26 Bends—Number in One Run. There shall not be more than the equivalent of four quarter bends (360 degrees total) between pull points, for example, conduit bodies and boxes.

358.28 Reaming and Threading.

(A) Reaming. All cut ends of EMT shall be reamed or otherwise finished to remove rough edges.

358.30 Securing and Supporting. EMT shall be installed as a complete system as provided in Article 300 and shall be securely fastened in place and supported in accordance with 358.30(A) and (B).

(A) Securely Fastened. EMT shall be securely fastened in place at least every 3 m (10 ft). In addition, each EMT run between termination points shall be securely fastened within 900 mm (3 ft) of each outlet box, junction box, device box, cabinet, conduit body, or other tubing termination.

Exception No. 1: Fastening of unbroken lengths shall be permitted to be increased to a distance of 1.5 m (5 ft) where structural members do not readily permit fastening within 900 mm (3 ft).

Exception No. 2: For concealed work in finished buildings or prefinished wall panels where such securing is impracticable, unbroken lengths (without coupling) of EMT shall be permitted to be fished.

(B) Supports. Horizontal runs of EMT supported by openings through framing members at intervals not greater than

3 m (10 ft) and securely fastened within 900 mm (3 ft) of termination points shall be permitted.

358.42 Couplings and Connectors. Couplings and connectors used with EMT shall be made up tight. Where buried in masonry or concrete, they shall be concretetight type. Where installed in wet locations, they shall be of the raintight type.

358.60 Grounding. EMT shall be permitted as an equipment grounding conductor.

ARTICLE 344
Rigid Metal Conduit: Type RMC

I. GENERAL

344.1 Scope. This article covers the use, installation, and construction specifications for rigid metal conduit (RMC) and associated fittings.

344.2 Definition

Rigid Metal Conduit (RMC). A threadable raceway of circular cross section designed for the physical protection and routing of conductors and cables and for use as an equipment grounding conductor when installed with its integral or associated coupling and appropriate fittings. RMC is generally made of steel (ferrous) with protective coatings or aluminum (nonferrous). Special use types are silicon bronze and stainless steel.

II. INSTALLATION

344.10 Uses Permitted.

(A) All Atmospheric Conditions and Occupancies. Use of RMC shall be permitted under all atmospheric conditions and occupancies. Ferrous raceways and fittings protected from corrosion solely by enamel shall be permitted only indoors and in occupancies not subject to severe corrosive influences.

(B) Corrosion Environments. RMC, elbows, couplings, and fittings shall be permitted to be installed in concrete, in direct contact with the earth, or in areas subject to severe corrosive influences where protected by corrosion protection and judged suitable for the condition.

(C) Cinder Fill. RMC shall be permitted to be installed in or under cinder fill where subject to permanent moisture where protected on all sides by a layer of noncinder concrete not less than 50 mm (2 in.) thick; where the conduit is not less than 450 mm (18 in.) under the fill; or where protected by corrosion protection and judged suitable for the condition.

(D) Wet Locations. All supports, bolts, straps, screws, and so forth, shall be of corrosion-resistant materials or protected against corrosion by corrosion-resistant materials.

344.24 Bends—How Made. Bends of RMC shall be made so that the conduit is not damaged and the internal diameter of the conduit is not effectively reduced. The radius of the curve of any field bend to the centerline of the conduit shall not be less than indicated in Table 344.24.

344.26 Bends—Number in One Run. There shall not be more than the equivalent of four quarter bends (360 degrees

Table 344.24 Radius of Conduit Bends

Conduit Size		One Shot and Full Shoe Benders		Other Bends	
Metric Designator	Trade Size	mm	in.	mm.	in.
16	1/2	101.6	4	101.6	4
21	3/4	114.3	4-1/2	127	5
27	1	146.05	5-3/4	152.4	6
35	1-1/4	184.15	7-1/4	203.2	8
41	1-1/2	209.55	8-1/4	254	10
53	2	241.3	9-1/2	304.8	12
63	2-1/2	266.7	10-1/2	381	15
78	3	330.2	13	457.2	18
91	3-1/2	381	15	533.4	21
103	4	406.4	16	609.6	24
129	5	609.6	24	762	30
155	6	762	30	914.4	36

total) between pull points, for example, conduit bodies and boxes.

344.28 Reaming and Threading. All cut ends shall be reamed or otherwise finished to remove rough edges. Where conduit is threaded in the field, a standard cutting die with a 1 in 16 taper (3/4-in. taper per foot) shall be used.

344.30 Securing and Supporting. RMC shall be installed as a complete system as provided in Article 300 and shall be securely fastened in place and supported in accordance with 344.30(A) and (B).

(A) Securely Fastened. RMC shall be securely fastened within 900 mm (3 ft) of each outlet box, junction box, device box, cabinet, conduit body, or other conduit termination. Fastening shall be permitted to be increased to a distance of 1.5 m (5 ft) where structural members do not readily permit fastening within 900 mm (3 ft). Where approved, conduit shall not be required to be securely fastened within 900 mm (3 ft) of the service head for above-the-roof termination of a mast.

(B) Supports. RMC shall be supported in accordance with one of the following.

(1) Conduit shall be supported at intervals not exceeding 3 m (10 ft).
(2) The distance between supports for straight runs of conduit shall be permitted in accordance with Table 346.30(B)(2), provided the conduit is made up with threaded couplings, and such supports prevent transmission of stresses to termination where conduit is deflected between supports.
(4) Horizontal runs of RMC supported by openings through framing members at intervals not exceeding 3 m (10 ft) and securely fastened within 900 mm (3 ft) of termination points shall be permitted.

344.42 Couplings and Connectors.

(A) Threadless. Threadless couplings and connectors used with conduit shall be made tight. Where buried in masonry or concrete, they shall be the concretetight type.

Table 344.30(B)(2) Supports for Rigid Metal Conduit

Conduit Size		Maximum Distance Between Rigid Metal Conduit Supports	
Metric Designator	Trade Size	m	ft
16–21	1/2–3/4	3.0	10
27	1	3.7	12
35–41	1-1/4–1-1/2	4.3	14
53–63	2–2-1/2	4.9	16
78 and larger	3 and larger	6.1	20

Where installed in wet locations, they shall be the raintight type. Threadless couplings and connectors shall not be used on threaded conduit ends unless listed for the purpose.

(B) Running Threads. Running threads shall not be used on conduit for connection at couplings.

344.46 Bushings. Where a conduit enters a box, fitting, or other enclosure, a bushing shall be provided to protect the wire from abrasion unless the design of the box, fitting, or enclosure is such as to afford equivalent protection.

FPN: See 300.4(F) for the protection of conductors sizes 4 AWG and larger at bushings.

344.60 Grounding. RMC shall be permitted as an equipment grounding conductor.

ARTICLE 352
Rigid Nonmetallic Conduit: Type RNC

I. GENERAL

352.1 Scope. This article covers the use, installation, and construction specifications for rigid nonmetallic conduit (RNC) and associated fittings.

352.2 Definition.

Rigid Nonmetallic Conduit (RNC). A nonmetallic raceway of circular cross section, with integral or associated couplings, connectors, and fittings for the installation of electrical conductors.

II. INSTALLATION

352.10 Uses Permitted. The use of RNC shall be permitted under the following conditions.

(A) Concealed. In walls, floors, and ceilings.

(B) Corrosive Influences. In locations subject to severe corrosive influences as covered in 300.6 and where subject to chemicals for which the materials are specifically approved.

(C) Cinders. In cinder fill.

(D) Wet Locations. In portions of dairies, laundries, canneries, or other wet locations and in locations where walls are frequently washed, the entire conduit system including boxes and fittings used therewith shall be installed and equipped so as to prevent water from entering the conduit. All supports, bolts, straps, screws, and so forth, shall be of corrosion-resistant materials or be protected against corrosion by approved corrosion-resistant materials.

(E) Dry and Damp Locations. In dry and damp locations not prohibited by 352.12.

(F) Exposed. For exposed work where not subject to physical damage if identified for such use.

(G) Underground Installations. For underground installations, see 300.5 and 300.50. Conduits listed for the purpose shall be permitted to be installed underground in continuous lengths from a reel.

(H) Support of Conduit Bodies. Rigid nonmetallic conduit shall be permitted to support nonmetallic conduit bodies not larger than the largest trade size of an entering raceway. The conduit bodies shall not contain devices or support luminaires (fixtures) or other equipment.

352.12 Uses Not Permitted. RNC shall not be used in the following locations.

(B) Support of Luminaires (Fixtures). For the support of luminaires (fixtures) or other equipment not described in 352.10(H).

(C) Physical Damage. Where subject to physical damage unless identified for such use.

(D) Ambient Temperatures. Where subject to ambient temperatures in excess of 50°C (122°F) unless listed otherwise.

(E) Insulation Temperature Limitations. For conductors whose insulation temperature limitations would exceed those for which the conduit is listed.

352.24 Bends—How Made. Bends shall be made so that the conduit will not be damaged and the internal diameter of the conduit will not be effectively reduced. Field bends shall be made only with bending equipment identified for the purpose. The radius of the curve to the centerline of such bends shall not be less than shown in Table 344.24, column "Other Bends."

352.26 Bends—Number in One Run. There shall not be more than the equivalent of four quarter bends (360 degrees total) between pull points, for example, conduit bodies and boxes.

352.28 Trimming. All cut ends shall be trimmed inside and outside to remove rough edges.

352.30 Securing and Supporting. RNC shall be installed as a complete system as provided in 300.18 and shall be fastened so that movement from thermal expansion or contraction is permitted. RNC shall be securely fastened and supported in accordance with 352.30(A) and (B).

(A) Securely Fastened. RNC shall be securely fastened within 900 mm (3 ft) of each outlet box, junction box, device box, conduit body, or other conduit termination. Conduit listed for securing at other than 900 mm (3 ft) shall be permitted to be installed in accordance with the listing.

(B) Supports. RNC shall be supported as required in Table 352.30(B). Conduit listed for support at spacings other

Table 352.30(B) Support of Rigid Nonmetallic Conduit (RNC)

Conduit Size		Maximum Spacing Between Supports	
Metric Designator	Trade Size	mm or m	ft
16–27	1/2–1	900 mm	3
35–53	1-1/4–2	1.5 m	5
63–78	2-1/2–3	1.8 m	6
91–129	3-1/2–5	2.1 m	7
155	6	2.5 m	8

than as shown in Table 352.30(B) shall be permitted to be installed in accordance with the listing. Horizontal runs of RNC supported by openings through framing members at intervals not exceeding those in Table 352.30(B) and securely fastened within 900 mm (3 ft) of termination points shall be permitted.

352.44 Expansion Fittings. Expansion fittings for RNC shall be provided to compensate for thermal expansion and contraction where the length change, in accordance with Table 352.44(A) or (B), is expected to be 6 mm (1/4 in.) or greater in a straight run between securely mounted items such as boxes, cabinets, elbows, or other conduit terminations.

352.46 Bushings. Where a conduit enters a box, fitting, or other enclosure, a bushing or adapter shall be provided to protect the wire from abrasion unless the box, fitting, or enclosure design provides equivalent protection.

FPN: See 300.4(F) for the protection of conductors 4 AWG and larger at bushings.

352.48 Joints. All joints between lengths of conduit, and between conduit and couplings, fittings, and boxes, shall be made by an approved method.

352.60 Grounding. Where equipment grounding is required by Article 250, a separate equipment grounding conductor shall be installed in the conduit.

Table 352.44(A) Expansion Characteristics of PVC Rigid Nonmetallic Conduit
Coefficient of Thermal Expansion = 6.084 × 10⁻⁵ mm/mm/°C (3.38 × 10⁻⁵ in./in./°F)

Temperature Change (°C)	Length Change of PVC Conduit (mm/m)	Temperature Change (°F)	Length Change of PVC Conduit (in./100 ft)	Temperature Change (°F)	Length Change of PVC Conduit (in./100 ft)
5	0.30	5	0.20	105	4.26
10	0.61	10	0.41	110	4.46
15	0.91	15	0.61	115	4.66
20	1.22	20	0.81	120	4.87
25	1.52	25	1.01	125	5.07
30	1.83	30	1.22	130	5.27
35	2.13	35	1.42	135	5.48
40	2.43	40	1.62	140	5.68
45	2.74	45	1.83	145	5.88
50	3.04	50	2.03	150	6.08
55	3.35	55	2.23	155	6.29
60	3.65	60	2.43	160	6.49
65	3.95	65	2.64	165	6.69
70	4.26	70	2.84	170	6.90
75	4.56	75	3.04	175	7.10
80	4.87	80	3.24	180	7.30
85	5.17	85	3.45	185	7.50
90	5.48	90	3.65	190	7.71
95	5.78	95	3.85	195	7.91
100	6.08	100	4.06	200	8.11

ARTICLE 362
Electrical Nonmetallic Tubing: Type ENT

I. GENERAL

362.1 Scope. This article covers the use, installation, and construction specifications for electrical nonmetallic tubing (ENT) and associated fittings.

362.2 Definition.

Electrical Nonmetallic Tubing (ENT). A nonmetallic pliable corrugated raceway of circular cross section with integral or associated couplings, connectors, and fittings for the installation of electric conductors. ENT is composed of a material that is resistant to moisture and chemical atmospheres and is flame retardant.

A pliable raceway is a raceway that can be bent by hand with a reasonable force, but without other assistance.

II. INSTALLATION

362.10 Uses Permitted. For the purpose of this article, the first floor of a building shall be that floor that has 50 percent or more of the exterior wall surface area level with or above finished grade. One additional level that is the first level and not designed for human habitation and used only for vehicle parking, storage, or similar use shall be permitted. The use of ENT and fittings shall be permitted in the following:

(1) In any building not exceeding three floors above grade

    a. For exposed work, where not prohibited by 362.12
    b. Concealed within walls, floors, and ceilings

(2) In any building exceeding three floors above grade, ENT shall be concealed within walls, floors, and ceilings where the walls, floors, and ceilings provide a thermal barrier of material that has at least a 15-minute finish rating as identified in listings of fire-rated assemblies. The 15-minute-finish-rated thermal barrier shall be permitted to be used for combustible walls, floors, and ceilings.

Exception: Where a fire sprinkler system(s) is installed in accordance with NFPA 13-1999, Standard for the Installa-

tion of Sprinkler Systems, on all floors, ENT is permitted to be used within walls, floors, and ceilings, exposed or concealed, in buildings exceeding three floors above grade.

(3) In locations subject to severe corrosive influences as covered in 300.6 and where subject to chemicals for which the materials are specifically approved.

(4) In concealed, dry, and damp locations not prohibited by 362.12.

(5) Above suspended ceilings where the suspended ceilings provide a thermal barrier of material that has at least a 15-minute finish rating as identified in listings of fire-rated assemblies, except as permitted in 362.10(1)(a).

Exception: Where a fire sprinkler system(s) is installed in accordance with NFPA 13-1999, Standard for the Installation of Sprinkler Systems, on all floors, ENT is permitted to be used within walls, floors, and ceilings, exposed or concealed, in buildings exceeding three floors above grade.

(6) Encased in poured concrete, or embedded in a concrete slab on grade where ENT is placed on sand or approved screenings, provided fittings identified for this purpose are used for connections.

(7) For wet locations indoors as permitted in this section or in a concrete slab on or below grade, with fittings listed for the purpose.

(8) Metric designator 16 through 27 (trade size 1/2 through 1) as listed manufactured prewired assembly.

362.12 Uses Not Permitted. ENT shall not be used in the following:

(2) For the support of luminaires (fixtures) and other equipment

(3) Where subject to ambient temperatures in excess of 50°C (122°F) unless listed otherwise

(4) For conductors whose insulation temperature limitations would exceed those for which the tubing is listed

(5) For direct earth burial

(7) In exposed locations, except as permitted by 362.10(1), 362.10(5), and 362.10(7)

(9) Where exposed to the direct rays of the sun, unless identified as sunlight resistant

(10) Where subject to physical damage

362.24 Bends—How Made. Bends shall be made so that the tubing will not be damaged and that the internal diameter of the tubing will not be effectively reduced. Bends shall be permitted to be made manually without auxiliary equipment, and the radius of the curve to the centerline of such bends shall not be less than shown in Table 344.24 using the column "Other Bends."

362.26 Bends—Number in One Run. There shall not be more than the equivalent of four quarter bends (360 degrees total) between pull points, for example, conduit bodies and boxes.

362.28 Trimming. All cut ends shall be trimmed inside and outside to remove rough edges.

362.30 Securing and Supporting. ENT shall be installed as a complete system as provided in Article 300 and shall be securely fastened in place and supported in accordance with 362.30(A) and (B).

(A) Securely Fastened. ENT shall be securely fastened at intervals not exceeding 900 mm (3 ft). In addition, ENT shall be securely fastened in place within 900 mm (3 ft) of each outlet box, device box, junction box, cabinet, or fitting where it terminates.

Exception: Lengths not exceeding a distance of 1.8 m (6 ft) from a luminaire (fixture) terminal connection for tap connections to lighting luminaires (fixtures) shall be permitted without being secured.

(B) Supports. Horizontal runs of ENT supported by openings in framing members at intervals not exceeding 900 mm (3 ft) and securely fastened within 900 mm (3 ft) of termination points shall be permitted.

362.46 Bushings. Where a tubing enters a box, fitting, or other enclosure, a bushing or adapter shall be provided to protect the wire from abrasion unless the box, fitting, or enclosure design provides equivalent protection.

FPN: See 300.4(F) for the protection of conductors size 4 AWG or larger.

362.48 Joints. All joints between lengths of tubing and between tubing and couplings, fittings, and boxes shall be by an approved method.

362.60 Grounding. Where equipment grounding is required by Article 250, a separate equipment grounding conductor shall be installed in the raceway.

ARTICLE 348
Flexible Metal Conduit: Type FMC

I. GENERAL

348.1 Scope. This article covers the use, installation, and construction specifications for flexible metal conduit (FMC) and associated fittings.

348.2 Definition.

Flexible Metal Conduit (FMC). A raceway of circular cross section made of helically wound, formed, interlocked metal strip.

II. INSTALLATION

348.10 Uses Permitted. FMC shall be permitted to be used in exposed and concealed locations.

348.12 Uses Not Permitted. FMC shall not be used in the following:

(1) In wet locations unless the conductors are approved for the specific conditions and the installation is such that liquid is not likely to enter raceways or enclosures to which the conduit is connected
(5) Where exposed to materials having a deteriorating effect on the installed conductors, such as oil or gasoline
(6) Underground or embedded in poured concrete or aggregate
(7) Where subject to physical damage

348.26 Bends—Number in One Run. There shall not be more than the equivalent of four quarter bends (360 degrees total) between pull points, for example, conduit bodies and boxes.

348.28 Trimming. All cut ends shall be trimmed or otherwise finished to remove rough edges, except where fittings that thread into the convolutions are used.

348.30 Securing and Supporting. FMC shall be securely fastened in place and supported in accordance with 348.30(A) and (B).

(A) Securely Fastened. FMC shall be securely fastened in place by an approved means within 300 mm (12 in.) of each box, cabinet, conduit body, or other conduit termination and shall be supported and secured at intervals not to exceed 1.4 m (4-1/2 ft).

Exception No. 1: Where FMC is fished.

Exception No. 2: Lengths not exceeding 900 mm (3 ft) at terminals where flexibility is required.

Exception No. 3: Lengths not exceeding 1.8 m (6 ft) from a luminaire (fixture) terminal connection for tap connections to luminaires (light fixtures) as permitted in 410.67(C).

(B) Supports. Horizontal runs of flexible metal conduit FMC supported by openings through framing members at intervals not greater than 1.4 m (4-1/2 ft) and securely fastened within 300 mm (12 in.) of termination points shall be permitted.

348.42 Couplings and Connectors. Angle connectors shall not be used for concealed raceway installations.

348.60 Grounding and Bonding. Where used to connect equipment where flexibility is required, an equipment grounding conductor shall be installed.

    Where required or installed, equipment grounding conductors shall be installed in accordance with 250.134(B).

    Where required or installed, equipment bonding jumpers shall be installed in accordance with 250.102.

ARTICLE 350
Liquidtight Flexible Metal Conduit: Type LFMC

I. GENERAL

350.1 Scope. This article covers the use, installation, and construction specifications for liquidtight flexible metal conduit (LFMC) and associated fittings.

350.2 Definition.

Liquidtight Flexible Metal Conduit (LFMC). A raceway of circular cross section having an outer liquidtight, nonmetallic, sunlight-resistant jacket over an inner flexible metal core with associated couplings, connectors, and fittings for the installation of electric conductors.

II. INSTALLATION

350.10 Uses Permitted. LFMC shall be permitted to be used in exposed or concealed locations as follows:

(1) Where conditions of installation, operation, or maintenance require flexibility or protection from liquids, vapors, or solids
(3) For direct burial where listed and marked for the purpose

350.12 Uses Not Permitted. LFMC shall not be used as follows:

(1) Where subject to physical damage
(2) Where any combination of ambient and conductor temperature produces an operating temperature in excess of that for which the material is approved

350.26 Bends—Number in One Run. There shall not be more than the equivalent of four quarter bends (360 degrees total) between pull points, for example, conduit bodies and boxes.

350.30 Securing and Supporting. LFMC shall be securely fastened in place and supported in accordance with 350.30(A) and (B).

(A) Securely Fastened. LFMC shall be securely fastened in place by an approved means within 300 mm (12 in.) of each box, cabinet, conduit body, or other conduit termination and shall be supported and secured at intervals not to exceed 1.4 m (4-1/2 ft).

Exception No. 1: Where LFMC is fished.

Exception No. 2: Lengths not exceeding 900 mm (3 ft) at terminals where flexibility is necessary.

Exception No. 3: Lengths not exceeding 1.8 m (6 ft) from a luminaire (fixture) terminal connection for tap conductors to luminaires (lighting fixtures), as permitted in 410.67(C).

(B) Supports. Horizontal runs of LFMC supported by openings through framing members at intervals not greater than 1.4 m (4-1/2 ft) and securely fastened within 300 mm (12 in.) of termination points shall be permitted.

350.42 Couplings and Connectors. Angle connectors shall not be used for concealed raceway installations.

350.60 Grounding and Bonding. Where used to connect equipment where flexibility is required, an equipment grounding conductor shall be installed.

Where required or installed, equipment grounding conductors shall be installed in accordance with 250.134(B).

Where required or installed, equipment bonding jumpers shall be installed in accordance with 250.102.

ARTICLE 356
Liquidtight Flexible Nonmetallic Conduit: Type LFNC

I. GENERAL

356.1 Scope. This article covers the use, installation, and construction specifications for liquidtight flexible nonmetallic conduit (LFNC) and associated fittings.

356.2 Definition.

Liquidtight Flexible Nonmetallic Conduit (LFNC). A raceway of circular cross section of various types as follows:

(1) A smooth seamless inner core and cover bonded together and having one or more reinforcement layers between the core and covers, designated as Type LFNC-A
(2) A smooth inner surface with integral reinforcement within the conduit wall, designated as Type LFNC-B
(3) A corrugated internal and external surface without integral reinforcement within the conduit wall, designated as LFNC-C.

II. INSTALLATION

356.10 Uses Permitted. LFNC shall be permitted to be used in exposed or concealed locations for the following purposes:

(1) Where flexibility is required for installation, operation, or maintenance
(2) Where protection of the contained conductors is required from vapors, liquids, or solids
(3) For outdoor locations where listed and marked as suitable for the purpose
(4) For direct burial where listed and marked for the purpose
(5) Type LFNC-B shall be permitted to be installed in lengths longer than 1.8 m (6 ft) where secured in accordance with 356.30

356.12 Uses Not Permitted. LFNC shall not be used as follows:

(1) Where subject to physical damage
(2) Where any combination of ambient and conductor temperatures is in excess of that for which the LFNC is approved

(3) In lengths longer than 1.8 m (6 ft), except as permitted by 356.100(5) or where a longer length is approved as essential for a required degree of flexibility

356.26 Bends—Number in One Run. There shall not be more than the equivalent of four quarter bends (360 degrees total) between pull points, for example, conduit bodies and boxes.

356.28 Trimming. All cut ends of conduit shall be trimmed inside and outside to remove rough edges.

356.30 Securing and Supporting. Type LFNC-B shall be securely fastened and supported in accordance with one of the following:

(1) The conduit shall be securely fastened at intervals not exceeding 900 mm (3 ft) and within 300 mm (12 in.) on each side of every outlet box, junction box, cabinet, or fitting.
(2) Securing or supporting of the conduit shall not be required where it is fished, installed in lengths not exceeding 900 mm (3 ft) at terminals where flexibility is required, or installed in lengths not exceeding 1.8 m (6 ft) from a luminaire (fixture) terminal connection for tap conductors to luminaires (lighting fixtures) permitted in 410.67(C).
(3) Horizontal runs of LFNC supported by openings through framing members at intervals not exceeding 900 mm (3 ft) and securely fastened within 300 mm (12 in.) of termination points shall be permitted.

356.42 Couplings and Connectors. Angle connectors shall not be used for concealed raceway installations.

356.60 Grounding and Bonding. Where used to connect equipment where flexibility is required, an equipment grounding conductor shall be installed.

Where required or installed, equipment grounding conductors shall be installed in accordance with 250.134(B).

Where required or installed, equipment bonding jumpers shall be installed in accordance with 250.102.

CHAPTER 15

WIRING METHODS AND CONDUCTORS

INTRODUCTION

This chapter contains provisions from Article 300—Wiring Methods and Article 310—Conductors for General Wiring. Article 300 covers wiring methods for all wiring installations unless modified by other articles. This article covers some general installation requirements for both cables and raceways. Some of the topics include: protection against physical damage; underground installations; securing and supporting; and boxes, conduit bodies, or fittings.

Article 310 covers general requirements for conductors, including their ampacity ratings. For dwelling unit applications, conductor ampacities specified in Table 310.16 are typically used. Under specified conditions in dwelling units, Table 310.15(B)(6) can be used to size service-entrance conductors, service lateral conductors, and feeder conductors that serve as the main power feeder. Conductors sized by Table 310.15(B)(6) must be fed from a 120/240-volt, 3-wire, single-phase source. These conductors can be installed in a raceway or cable, with or without an equipment grounding conductor.

ARTICLE 300
Wiring Methods and Conductors

I. GENERAL REQUIREMENTS

300.1 Scope.

(A) All Wiring Installations. This article covers wiring methods for all wiring installations unless modified by other articles.

300.4 Protection Against Physical Damage. Where subject to physical damage, conductors shall be adequately protected.

(A) Cables and Raceways Through Wood Members.

(1) Bored Holes. In both exposed and concealed locations, where a cable- or raceway-type wiring method is installed through bored holes in joists, rafters, or wood members, holes shall be bored so that the edge of the hole is not less than 32 mm (1-1/4 in.) from the nearest edge of the wood member. Where this distance cannot be maintained, the cable or raceway shall be protected from penetration by screws or nails by a steel plate or bushing, at least 1.6 mm (1/16 in.) thick, and of appropriate length and width installed to cover the area of the wiring.

Exception: Steel plates shall not be required to protect rigid metal conduit, intermediate metal conduit, rigid non-metallic conduit, or electrical metallic tubing.

(2) Notches in Wood. Where there is no objection because of weakening the building structure, in both exposed and concealed locations, cables or raceways shall be permitted to be laid in notches in wood studs, joists, rafters, or other wood members where the cable or raceway at those points is protected against nails or screws by a steel plate at least 1.6 mm (1/16 in.) thick installed before the building finish is applied.

Exception: Steel plates shall not be required to protect rigid metal conduit, intermediate metal conduit, rigid non-metallic conduit, or electrical metallic tubing.

(B) Nonmetallic-Sheathed Cables and Electrical Nonmetallic Tubing Through Metal Framing Members.

(1) Nonmetallic-Sheathed Cable. In both exposed and concealed locations where nonmetallic-sheathed cables pass through either factory or field punched, cut, or drilled slots or holes in metal members, the cable shall be protected by listed bushings or listed grommets covering all metal edges that are securely fastened in the opening prior to installation of the cable.

(2) Nonmetallic-Sheathed Cable and Electrical Nonmetallic Tubing. Where nails or screws are likely to penetrate nonmetallic-sheathed cable or electrical nonmetallic tubing, a steel sleeve, steel plate, or steel clip not less than 1.6 mm (1/16 in.) in thickness shall be used to protect the cable or tubing.

(C) Cables Through Spaces Behind Panels Designed to Allow Access. Cables or raceway-type wiring methods, installed behind panels designed to allow access, shall be supported according to their applicable articles.

(D) Cables and Raceways Parallel to Framing Members. In both exposed and concealed locations, where a cable- or raceway-type wiring method is installed parallel to framing members, such as joists, rafters, or studs, the cable or raceway shall be installed and supported so that the nearest outside surface of the cable or raceway is not less than 32 mm (1-1/4 in.) from the nearest edge of the framing member where nails or screws are likely to penetrate. Where this distance cannot be maintained, the cable or raceway shall be protected from penetration by nails or screws by a steel plate, sleeve, or equivalent at least 1.6 mm (1/16 in.) thick.

Exception No. 1: Steel plates, sleeves, or the equivalent shall not be required to protect rigid metal conduit, intermediate metal conduit, rigid nonmetallic conduit, or electrical metallic tubing.

Exception No. 2: For concealed work in finished buildings, or finished panels for prefabricated buildings where such supporting is impracticable, it shall be permissible to fish the cables between access points.

(E) Cables and Raceways Installed in Shallow Grooves. Cable- or raceway-type wiring methods installed in a groove, to be covered by wallboard, siding, paneling, carpeting, or similar finish, shall be protected by 1.6 mm (1/16 in.) thick steel plate, sleeve, or equivalent or by not less than 32 mm (1-1/4 in.) free space for the full length of the groove in which the cable or raceway is installed.

(F) Insulated Fittings. Where raceways containing ungrounded conductors 4 AWG or larger enter a cabinet, box enclosure, or raceway, the conductors shall be protected by a substantial fitting providing a smoothly rounded insulating surface, unless the conductors are separated from the fitting or raceway by substantial insulating material that is securely fastened in place.

Exception: Where threaded hubs or bosses that are an integral part of a cabinet, box enclosure, or raceway provide a smoothly rounded or flared entry for conductors.

Conduit bushings constructed wholly of insulating material shall not be used to secure a fitting or raceway. The insulating fitting or insulating material shall have a temperature rating not less than the insulation temperature rating of the installed conductors.

300.5 Underground Installations.

(A) Minimum Cover Requirements. Direct-buried cable or conduit or other raceways shall be installed to meet the minimum cover requirements of Table 300.5.

(B) Grounding. All underground installations shall be grounded and bonded in accordance with Article 250.

(C) Underground Cables Under Buildings. Underground cable installed under a building shall be in a raceway that is extended beyond the outside walls of the building.

(D) Protection from Damage. Direct-buried conductors and cables shall be protected from damage in accordance with (1) through (5).

Table 300.5 Minimum Cover Requirements, 0 to 600 Volts, Nominal, Burial in Millimeters (Inches)

Location of Wiring Method or Circuit	Column 1 Direct Burial Cables or Conductors		Column 2 Rigid Metal Conduit or Intermediate Metal Conduit		Column 3 Nonmetallic Raceways Listed for Direct Burial Without Concrete Encasement or Other Approved Raceways		Column 4 Residential Branch Circuits Rated 120 Volts or Less with GFCI Protection and Maximum Overcurrent Protection of 20 Amperes		Column 5 Circuits for Control of Irrigation and Landscape Lighting Limited to Not More than 30 Volts and Installed with Type UF or in Other Identified Cable or Raceway	
	mm	in.	mm	in.	mm	in.	mm	in.	mm	in.
All locations not specified below	600	24	150	6	450	18	300	12	150	6
In trench below 50-mm (2-in.) thick concrete or equivalent	450	18	150	6	300	12	50	6	50	6
Under a building	0 (in raceway only)	0	0	0	0	0	0 (in raceway only)	0	0 (in raceway only)	0

Continued

171

Table 300.5 Minimum Cover Requirements, 0 to 600 Volts, Nominal, Burial in Millimeters (Inches) *Continued*

				Type of Wiring Method or Circuit					
	Column 1 Direct Burial Cables or Conductors		Column 2 Rigid Metal Conduit or Intermediate Metal Conduit		Column 3 Nonmetallic Raceways Listed for Direct Burial Without Concrete Encasement or Other Approved Raceways		Column 4 Residential Branch Circuits Rated 120 Volts or Less with GFCI Protection and Maximum Overcurrent Protection of 20 Amperes		Column 5 Circuits for Control of Irrigation and Landscape Lighting Limited to Not More than 30 Volts and Installed with Type UF or in Other Identified Cable or Raceway
Location of Wiring Method or Circuit	mm	in.	mm	in.	mm	in.	mm	in.	mm	in.
Under minimum of 102 mm. (4-in.) thick concrete exterior slab with no vehicular traffic and the slab extending not less than 152 mm (6 in.) beyond the underground installation	450	18	100	4	100	4	150 (direct burial) 100 (in raceway)	6 4	150 (direct burial)	6
Under streets, highways, roads, alleys, driveways, and parking lots	600	24	600	24	600	24	600	24	600	24

One- and two-family dwelling driveways and outdoor parking areas, and used only for dwelling-related purposes	450	18	450	18	450	18	300	12	450	18
In or under airport runways, including adjacent areas where trespassing prohibited	450	18	450	18	450	18	450	18	450	18

Notes:
1. Cover is defined as the shortest distance in millimeters (inches) measured between a point on the top surface of any direct-buried conductor, cable, conduit, or other raceway and the top surface of finished grade, concrete, or similar cover.
2. Raceways approved for burial only where concrete encased shall require concrete envelope not less than 50 mm (2 in.) thick.
3. Lesser depths shall be permitted where cables and conductors rise for terminations or splices or where access is otherwise required.
4. Where one of the wiring method types listed in Columns 1–3 is used for one of the circuit types in Columns 4 and 5, the shallower depth of burial shall be permitted.
5. Where solid rock prevents compliance with the cover depths specified in this table, the wiring shall be installed in metal or nonmetallic raceway permitted for direct burial. The raceways shall be covered by a minimum of 50 mm (2 in.) of concrete extending down to rock.

173

(1) Emerging from Grade. Direct-buried conductors and enclosures emerging from grade shall be protected by enclosures or raceways extending from the minimum cover distance required by 300.5(A) below grade to a point at least 2.5 m (8 ft) above finished grade. In no case shall the protection be required to exceed 450 mm (18 in.) below finished grade.

(2) Conductors Entering Buildings. Conductors entering a building shall be protected to the point of entrance.

(3) Service Conductors. Underground service conductors that are not encased in concrete and that are buried 450 mm (18 in.) or more below grade shall have their location identified by a warning ribbon that is placed in the trench at least 300 mm (12 in.) above the underground installation.

(4) Enclosure or Raceway Damage. Where the enclosure or raceway is subject to physical damage, the conductors shall be installed in rigid metal conduit, intermediate metal conduit, Schedule 80 rigid nonmetallic conduit, or equivalent.

(5) Listing. Cables and insulated conductors installed in enclosures or raceways in underground installations shall be listed for use in wet locations.

(E) Splices and Taps. Direct-buried conductors or cables shall be permitted to be spliced or tapped without the use of splice boxes. The splices or taps shall be made in accordance with 110.14(B).

(F) Backfill. Backfill that contains large rocks, paving materials, cinders, large or sharply angular substances, or corrosive material shall not be placed in an excavation where materials may damage raceways, cables, or other substructures or prevent adequate compaction of fill or contribute to corrosion of raceways, cables, or other substructures.

Where necessary to prevent physical damage to the raceway or cable, protection shall be provided in the form of granular or selected material, suitable running boards, suitable sleeves, or other approved means.

(G) Raceway Seals. Conduits or raceways through which moisture may contact energized live parts shall be sealed or plugged at either or both ends.

(H) Bushing. A bushing, or terminal fitting, with an integral bushed opening shall be used at the end of a conduit or other raceway that terminates underground where the conductors or cables emerge as a direct burial wiring method. A seal incorporating the physical protection characteristics of a bushing shall be permitted to be used in lieu of a bushing.

(I) Conductors of the Same Circuit. All conductors of the same circuit and, where used, the grounded conductor and all equipment grounding conductors shall be installed in the same raceway or cable or shall be installed in close proximity in the same trench.

Exception No. 1: Conductors in parallel in raceways or cables shall be permitted, but each raceway or cable shall contain all conductors of the same circuit including grounding conductors.

(J) Ground Movement. Where direct-buried conductors, raceways, or cables are subject to movement by settlement or frost, direct-buried conductors, raceways, or cables shall be arranged to prevent damage to the enclosed conductors or to equipment connected to the raceways.

300.10 Electrical Continuity of Metal Raceways and Enclosures. Metal raceways, cable armor, and other metal enclosures for conductors shall be metallically joined together into a continuous electric conductor and shall be connected to all boxes, fittings, and cabinets so as to provide effective electrical continuity. Unless specifically permitted elsewhere in this *Code*, raceways and cable assemblies shall be mechanically secured to boxes, fittings, cabinets, and other enclosures.

Exception No. 1: Short sections of raceways used to provide support or protection of cable assemblies from physical damage shall not be required to be made electrically continuous.

300.11 Securing and Supporting.

(A) Secured in Place. Raceways, cable assemblies, boxes, cabinets, and fittings shall be securely fastened in place. Support wires that do not provide secure support shall not be permitted as the sole support. Support wires and associated fittings that provide secure support and that are installed in addition to the ceiling grid support wires shall be permitted as the sole support. Where independent support wires are used, they shall be secured at both ends. Cables and raceways shall not be supported by ceiling grids.

(1) Fire-Rated Assemblies. Wiring located within the cavity of a fire-rated floor–ceiling or roof–ceiling assembly shall not be secured to, or supported by, the ceiling assembly, including the ceiling support wires. An independent means of secure support shall be provided. Where independent support wires are used, they shall be distinguishable by color, tagging, or other effective means from those that are part of the fire-rated design.

Exception: The ceiling support system shall be permitted to support wiring and equipment that have been tested as part of the fire-rated assembly.

(2) Non–Fire-Rated Assemblies. Wiring located within the cavity of a non–fire-rated floor–ceiling or roof–ceiling assembly shall not be secured to, or supported by, the ceiling assembly, including the ceiling support wires. An independent means of secure support shall be provided.

Exception: The ceiling support system shall be permitted to support branch-circuit wiring and associated equipment where installed in accordance with the ceiling system manufacturer's instructions.

(B) Raceways Used as Means of Support. Raceways shall only be used as a means of support for other raceways, cables, or nonelectric equipment under the following conditions:

(1) Where the raceway or means of support is identified for the purpose; or
(2) Where the raceway contains power supply conductors for electrically controlled equipment and is used to support Class 2 circuit conductors or cables that are solely for the purpose of connection to the equipment control circuits; or

(3) Where the raceway is used to support boxes or conduit bodies in accordance with 314.23 or to support luminaires (fixtures) in accordance with 410.16(F)

(C) Cables Not Used as Means of Support. Cable wiring methods shall not be used as a means of support for other cables, raceways, or nonelectrical equipment.

300.12 Mechanical Continuity—Raceways and Cables. Metal or nonmetallic raceways, cable armors, and cable sheaths shall be continuous between cabinets, boxes, fittings, or other enclosures or outlets.

Exception: Short sections of raceways used to provide support or protection of cable assemblies from physical damage shall not be required to be mechanically continuous.

300.13 Mechanical and Electrical Continuity—Conductors.

(A) General. Conductors in raceways shall be continuous between outlets, boxes, devices, and so forth. There shall be no splice or tap within a raceway unless permitted by 300.15; 368.8(A); 376.56; 378.56; 384.56; 386.56; 388.56; or 390.6.

(B) Device Removal. In multiwire branch circuits, the continuity of a grounded conductor shall not depend on device connections such as lampholders, receptacles, and so forth, where the removal of such devices would interrupt the continuity.

300.14 Length of Free Conductors at Outlets, Junctions, and Switch Points. At least 150 mm (6 in.) of free conductor, measured from the point in the box where it emerges from its raceway or cable sheath, shall be left at each outlet, junction, and switch point for splices or the connection of luminaires (fixtures) or devices. Where the opening to an outlet, junction, or switch point is less than 200 mm (8 in.) in any dimension, each conductor shall be long enough to extend at least 75 mm (3 in.) outside the opening.

Exception: Conductors that are not spliced or terminated at the outlet, junction, or switch point shall not be required to comply with 300.14.

300.15 Boxes, Conduit Bodies, or Fittings—Where Required. A box shall be installed at each outlet and switch point for concealed knob-and-tube wiring.

Fittings and connectors shall be used only with the specific wiring methods for which they are designed and listed.

Where the wiring method is conduit, tubing, Type AC cable, Type MC cable, Type MI cable, nonmetallic-sheathed cable, or other cables, a box or conduit body complying with Article 314 shall be installed at each conductor splice point, outlet point, switch point, junction point, termination point, or pull point, unless otherwise permitted in 300.15(A) through (M).

(A) Wiring Methods with Interior Access. A box or conduit body shall not be required for each splice, junction, switch, pull, termination, or outlet points in wiring methods with removable covers, such as wireways, multioutlet assemblies, auxiliary gutters, and surface raceways. The covers shall be accessible after installation.

(B) Equipment. An integral junction box or wiring compartment as part of approved equipment shall be permitted in lieu of a box.

(C) Protection. A box or conduit body shall not be required where cables enter or exit from conduit or tubing that is used to provide cable support or protection against physical damage. A fitting shall be provided on the end(s) of the conduit or tubing to protect the cable from abrasion.

(F) Fitting. A fitting identified for the use shall be permitted in lieu of a box or conduit body where conductors are not spliced or terminated within the fitting. The fitting shall be accessible after installation.

(G) Direct-Buried Conductors. As permitted in 300.5(E), a box or conduit body shall not be required for splices and taps in direct-buried conductors and cables.

(H) Insulated Devices. As permitted in 334.40(B), a box or conduit body shall not be required for insulated devices supplied by nonmetallic-sheathed cable.

300.16 Raceway or Cable to Open or Concealed Wiring.

(A) Box or Fitting. A box or terminal fitting having a separately bushed hole for each conductor shall be used wherever a change is made from conduit, electrical metallic tubing, electrical nonmetallic tubing, nonmetallic-sheathed cable, Type AC cable, Type MC cable, or mineral-insulated, metal-sheathed cable and surface raceway wiring to open wiring or to concealed knob-and-tube wiring. A fitting used for this purpose shall contain no taps or splices and shall not be used at luminaire (fixture) outlets.

(B) Bushing. A bushing shall be permitted in lieu of a box or terminal where the conductors emerge from a raceway and enter or terminate at equipment, such as open switchboards, unenclosed control equipment, or similar equipment. The bushing shall be of the insulating type for other than lead-sheathed conductors.

300.18 Raceway Installations.

(A) Complete Runs. Raceways, other than busways or exposed raceways having hinged or removable covers, shall be installed complete between outlet, junction, or splicing points prior to the installation of conductors. Where required to facilitate the installation of utilization equipment, the raceway shall be permitted to be initially installed without a terminating connection at the equipment. Prewired raceway assemblies shall be permitted only where specifically permitted in this *Code* for the applicable wiring method.

300.21 Spread of Fire or Products of Combustion. Electrical installations in hollow spaces, vertical shafts, and ventilation or air-handling ducts shall be made so that the possible spread of fire or products of combustion will not be substantially increased. Openings around electrical penetrations through fire-resistant-rated walls, partitions, floors, or ceilings shall be firestopped using approved methods to maintain the fire resistance rating.

300.22 Wiring in Ducts, Plenums, and Other Air-Handling Spaces. The provisions of this section apply to the installation and uses of electric wiring and equipment in ducts, plenums, and other air-handling spaces.

(C) Other Space Used for Environmental Air. This section applies to space used for environmental air-handling purposes other than ducts and plenums as specified in 300.22(A) and (B). It does not include habitable rooms or areas of buildings, the prime purpose of which is not air handling.

Exception: This section shall not apply to the joist or stud spaces of dwelling units where the wiring passes through such spaces perpendicular to the long dimension of such spaces.

ARTICLE 310
Conductors for General Wiring

310.1 Scope. This article covers general requirements for conductors and their type designations, insulations, markings, mechanical strengths, ampacity ratings, and uses. These requirements do not apply to conductors that form an integral part of equipment, such as motors, motor controllers, and similar equipment, or to conductors specifically provided for elsewhere in this *Code*.

310.2 Conductors.

(A) Insulated. Conductors shall be insulated.

Exception: Where covered or bare conductors are specifically permitted elsewhere in this Code.

(B) Conductor Material. Conductors in this article shall be of aluminum, copper-clad aluminum, or copper unless otherwise specified.

310.3 Stranded Conductors. Where installed in raceways, conductors of size 8 AWG and larger shall be stranded.

310.4 Conductors in Parallel. Aluminum, copper-clad aluminum, or copper conductors of size 1/0 AWG and larger, comprising each phase, neutral, or grounded circuit conductor, shall be permitted to be connected in parallel (electrically joined at both ends to form a single conductor).

The paralleled conductors in each phase, neutral, or grounded circuit conductor shall

(1) Be the same length
(2) Have the same conductor material
(3) Be the same size in circular mil area
(4) Have the same insulation type
(5) Be terminated in the same manner

Where run in separate raceways or cables, the raceways or cables shall have the same physical characteristics. Conductors of one phase, neutral, or grounded circuit conductor shall not be required to have the same physical characteristics as those of another phase, neutral, or grounded circuit conductor to achieve balance.

Where equipment grounding conductors are used with conductors in parallel, they shall comply with the requirements of this section except that they shall be sized in accordance with 250.122.

Conductors installed in parallel shall comply with the provisions of 310.15(B)(2)(a).

310.7 Direct Burial Conductors. Conductors used for direct burial applications shall be of a type identified for such use.

Cables rated above 2000 volts shall be shielded.

The metallic shield, sheath, or armor shall be grounded through an effective grounding path meeting the requirements of 250.4(A)(5) or 250.4(B)(4).

310.8 Locations.

(A) Dry Locations. Insulated conductors and cables used in dry locations shall be any of the types identified in this *Code*.

(B) Dry and Damp Locations. Insulated conductors and cables used in dry and damp locations shall be Types FEP, FEPB, MTW, PFA, RHH, RHW, RHW-2, SA, THHN, THW, THW-2, THHW, THHW-2, THWN, THWN-2, TW, XHH, XHHW, XHHW-2, Z, or ZW.

(C) Wet Locations. Insulated conductors and cables used in wet locations shall be

(1) Moisture-impervious metal-sheathed;
(2) Types MTW, RHW, RHW-2, TW, THW, THW-2, THHW, THHW-2, THWN, THWN-2, XHHW, XHHW-2, ZW; or
(3) Of a type listed for use in wet locations.

(D) Locations Exposed to Direct Sunlight. Insulated conductors and cables used where exposed to direct rays of the sun shall be of a type listed for sunlight resistance or listed and marked "sunlight resistant."

310.15 Ampacities for Conductors Rated 0–2000 Volts.

(B) Tables. Ampacities for conductors rated 0 to 2000 volts shall be as specified in the Allowable Ampacity Tables 310.16 through 310.19 and Ampacity Tables 310.20 through 310.23 as modified by (1) through (6).

> FPN: Tables 310.16 through 310.19 are application tables for use in determining conductor sizes on loads calculated in accordance with Article 220. Allowable ampacities result from consideration of one or more of the following:
>
> (1) Temperature compatibility with connected equipment, especially the connection points.
> (2) Coordination with circuit and system overcurrent protection.
> (3) Compliance with the requirements of product listings or certifications. See 110.3(B).
> (4) Preservation of the safety benefits of established industry practices and standardized procedures.

(2) Adjustment Factors.

(a) More Than Three Current-Carrying Conductors in a Raceway or Cable. Where the number of current-carrying conductors in a raceway or cable exceeds three, or where single conductors or multiconductor cables are stacked or bundled longer than 600 mm (24 in.) without maintaining spacing and are not installed in raceways, the allowable ampacity of each conductor shall be reduced as shown in Table 310.15(B)(2)(a).

Table 310.15(B)(2)(a) Adjustment Factors for More Than Three Current-Carrying Conductors in a Raceway or Cable

Number of Current-Carrying Conductors	Percent of Values in Tables 310.16 through 310.19 as Adjusted for Ambient Temperature if Necessary
4–6	80
7–9	70
10–20	50
21–30	45
31–40	40
41 and above	35

Exception No. 3: Derating factors shall not apply to conductors in nipples having a length not exceeding 600 mm (24 in.).

Exception No. 4: Derating factors shall not apply to underground conductors entering or leaving an outdoor trench if those conductors have physical protection in the form of rigid metal conduit, intermediate metal conduit, or rigid nonmetallic conduit having a length not exceeding 3.05 m (10 ft) and if the number of conductors does not exceed four.

(3) Bare or Covered Conductors. Where bare or covered conductors are used with insulated conductors, their allowable ampacities shall be limited to those permitted for the adjacent insulated conductors.

(4) Neutral Conductor.

(a) A neutral conductor that carries only the unbalanced current from other conductors of the same circuit shall not be required to be counted when applying the provisions of 310.15(B)(2)(a).

(5) Grounding or Bonding Conductor. A grounding or bonding conductor shall not be counted when applying the provisions of 310.15(B)(2)(a).d

(6) 120/240-Volt, 3-Wire, Single-Phase Dwelling Services and Feeders. For dwelling units, conductors, as listed in Table 310.15(B)(6), shall be permitted as 120/240-volt, 3-wire,

Table 310.15(B)(6) Conductor Types and Sizes for 120/240-Volt, 3-Wire, Single-Phase Dwelling Services and Feeders

Conductor (AWG or kcmil)		
Copper	Aluminum or Copper-Clad Aluminum	Service or Feeder Rating (Amperes)
4	2	100
3	1	110
2	1/0	125
1	2/0	150
1/0	3/0	175
2/0	4/0	200
3/0	250	225
4/0	300	250
250	350	300
350	500	350
400	600	400

single-phase service-entrance conductors, service lateral conductors, and feeder conductors that serve as the main power feeder to a dwelling unit and are installed in raceway or cable with or without an equipment grounding conductor. For application of this section, the main power feeder shall be the feeder(s) between the main disconnect and the lighting and appliance branch-circuit panelboard(s). The feeder conductors to a dwelling unit shall not be required to be larger than their service-entrance conductors. The grounded conductor shall be permitted to be smaller than the ungrounded conductors, provide the requirements of 215.2, 220.22, and 230.42 are met.

Table 310.16 Allowable Ampacities of Insulated Conductors Rated 0 Through 2000 Volts, 60°C Through 90°C (140°F Through 194°F), Not More than Three Current-Carrying Conductors in Raceway, Cable, or Earth (Directly Buried), Based on Ambient Temperature of 30°C (86°F)

Size AWG or kcmil	Temperature Rating of Conductor (See Table 310.13)						Size AWG or kcmil
	60°C (140°F)	75°C (167°F)	90°C (194°F)	60°C (140°F)	75°C (167°F)	90°C (194°F)	
	Types TW, UF	Types RHW, THHW, THW, THWN, XHHW, USE, ZW	Types TBS, SA, SIS, FEP, FEPB, MI, RHH, RHW-2, THHN, THHW, THW-2, THWN-2, USE-2, XHH, XHHW, XHHW-2, ZW-2	Types TW, UF	Types RHW, THHW, THW, THWN, XHHW, USE	Types TBS, SA, SIS, THHN, THHW, THW-2, THWN-2, RHH, RHW-2, USE-2, XHH, XHHW, XHHW-2, ZW-2	
	Copper			Aluminum or Copper-Clad Aluminum			
18	—	—	14	—	—	—	18
16	—	—	18	—	—	—	16
14*	20	20	25	—	—	—	14*
12*	25	25	30	20	20	25	12*
10*	30	35	40	25	30	35	10*
8	40	50	55	30	40	45	8
6	55	65	75	40	50	60	6

Continued

Table 310.16 Allowable Ampacities of Insulated Conductors Rated 0 Through 2000 Volts, 60°C Through 90°C (140°F Through 194°F), Not More than Three Current-Carrying Conductors in Raceway, Cable, or Earth (Directly Buried), Based on Ambient Temperature of 30°C (86°F) *Continued*

Size AWG or kcmil	Temperature Rating of Conductor (See Table 310.13)						Size AWG or kcmil
	60°C (140°F)	75°C (167°F)	90°C (194°F)	60°C (140°F)	75°C (167°F)	90°C (194°F)	
	Types TW, UF	Types RHW, THHW, THW, THWN, XHHW, USE, ZW	Types TBS, SA, SIS, FEP, FEPB, MI, RHH, RHW-2, THHN, THHW, THW-2, THWN-2, USE-2, XHH, XHHW, XHHW-2, ZW-2	Types TW, UF	Types RHW, THW, THHW, THWN, USE	Types TBS, SA, SIS, THHN, THHW, THW-2, THWN-2, RHH, RHW-2, USE-2, XHH, XHHW, XHHW-2, ZW-2	
	Copper			Aluminum or Copper-Clad Aluminum			
4	70	85	95	55	65	75	4
3	85	100	110	65	75	85	3
2	95	115	130	75	90	100	2
1	110	130	150	85	100	115	1
1/0	125	150	170	100	120	135	1/0
2/0	145	175	195	115	135	150	2/0
3/0	165	200	225	130	155	175	3/0
4/0	195	230	260	150	180	205	4/0

250	215	255	290	170	205	230	250
300	240	285	320	190	230	255	300
350	260	310	350	210	250	280	350
400	280	335	380	225	270	305	400
500	320	380	430	260	310	350	500
600	355	420	475	285	340	385	600
700	385	460	520	310	375	420	700
750	400	475	535	320	385	435	750
800	410	490	555	330	395	450	800
900	435	520	585	355	425	480	900
1000	455	545	615	375	445	500	1000
1250	495	590	665	405	485	545	1250
1500	520	625	705	435	520	585	1500
1750	545	650	735	455	545	615	1750
2000	560	665	750	470	560	630	2000

Continued

Table 310.16 Allowable Ampacities of Insulated Conductors Rated 0 Through 2000 Volts, 60°C Through 90°C (140°F Through 194°F), Not More than Three Current-Carrying Conductors in Raceway, Cable, or Earth (Directly Buried), Based on Ambient Temperature of 30°C (86°F) *Continued*

Size	Temperature Rating of Conductor (See Table 310.13)					Size
	60°C (140°F)	75°C (167°F)	90°C (194°F)	75°C (167°F)	90°C (194°F)	
AWG or kcmil	Types TW, UF	Types RHW, THW, THWN, THHW, USE, ZW	Types TBS, SA, SIS, FEP, FEPB, MI, RHH, RHW-2, THHN, THHW, THW-2, THWN-2, USE-2, XHH, XHHW, XHHW-2, ZW-2	Types RHW, THW, THHW, THWN, XHHW, USE	Types TBS, SA, SIS, THHN, THHW, THW-2, THWN-2, RHH, RHW-2, USE-2, XHH, XHHW, XHHW-2, ZW-2	AWG or kcmil

CORRECTION FACTORS

For ambient temperatures other than 30°C (86°F), multiply the allowable ampacities shown above by the appropriate factor shown below.

Ambient Temp. (°C)						Ambient Temp. (°F)
21–25	1.08	1.05	1.04	1.05	1.04	70–77
26–30	1.00	1.00	1.00	1.00	1.00	78–86
31–35	0.91	0.94	0.96	0.94	0.96	87–95
36–40	0.82	0.88	0.91	0.88	0.91	96–104
41–45	0.71	0.82	0.87	0.82	0.87	105–113
46–50	0.58	0.75	0.82	0.75	0.82	114–122
51–55	0.41	0.67	0.76	0.67	0.76	123–131
56–60	—	0.58	0.71	0.58	0.71	132–140
61–70	—	0.33	0.58	0.33	0.58	141–158
71–80	—	—	0.41	—	0.41	159–176

*See 240.4(D)

CHAPTER 16

BRANCH CIRCUITS

INTRODUCTION

This chapter contains provisions from Article 210—Branch Circuits. The provisions within Article 210 have been divided into three chapters of this pocket guide. This chapter covers Part I (general provisions) and Part II (branch-circuit ratings). Some of the general provisions pertain to multi-wire branch circuits, ground-fault circuit-interrupters, required branch circuits, and arc-fault circuit interrupters. Part II covers the branch-circuit ratings for conductors and receptacles.

Chapter 17 contains the receptacle outlet provisions within Article 210, and Chapter 19 contains the lighting outlet provisions.

ARTICLE 210
Branch Circuits

I. GENERAL PROVISIONS

210.1 Scope. This article covers branch circuits except for branch circuits that supply only motor loads, which are covered in Article 430. Provisions of this article and Article 430 apply to branch circuits with combination loads.

210.3 Rating. Branch circuits recognized by this article shall be rated in accordance with the maximum permitted ampere rating or setting of the overcurrent device. The rating for other than individual branch circuits shall be 15, 20, 30, 40, and 50 amperes. Where conductors of higher ampacity are used for any reason, the ampere rating or setting of the specified overcurrent device shall determine the circuit rating.

210.4 Multiwire Branch Circuits.

(A) General. Branch circuits recognized by this article shall be permitted to be multiwire circuits. A multiwire branch circuit shall be permitted to be considered as multiple circuits. All conductors shall originate from the same panelboard.

(B) Dwelling Units. In dwelling units, a multiwire branch circuit supplying more than one device or equipment on the same yoke shall be provided with a means to disconnect simultaneously all ungrounded conductors at the panelboard where the branch circuit originated.

(C) Line-to-Neutral Loads. Multiwire branch circuits shall supply only line-to-neutral loads.

Exception No. 1: A multiwire branch circuit that supplies only one utilization equipment.

Exception No. 2: Where all ungrounded conductors of the multiwire branch circuit are opened simultaneously by the branch-circuit overcurrent device.

210.5 Identification for Branch Circuits.

(A) Grounded Conductor. The grounded conductor of a branch circuit shall be identified in accordance with 200.6.

(B) Equipment Grounding Conductor. The equipment grounding conductor shall be identified in accordance with 250.119.

210.7 Branch Circuit Receptacle Requirements.

(A) Receptacle Outlet Location. Receptacle outlets shall be located in branch circuits in accordance with Part III of Article 210.

(B) Receptacle Requirements. Specific requirements for receptacles are covered in Article 406.

(C) Multiple Branch Circuits. Where more than one branch circuit supplies more than one receptacle on the same yoke, a means to simultaneously disconnect the ungrounded conductors supplying those receptacles shall be provided at the panelboard where the branch circuits originated.

210.8 Ground-Fault Circuit-Interrupter Protection for Personnel.

(A) Dwelling Units. All 125-volt, single-phase, 15- and 20-ampere receptacles installed in the locations specified in (1) through (8) shall have ground-fault circuit-interrupter protection for personnel.

(1) Bathrooms
(2) Garages, and also accessory buildings that have a floor located at or below grade level not intended as habitable rooms and limited to storage areas, work areas, and areas of similar use

Exception No. 1: Receptacles that are not readily accessible.

Exception No. 2: A single receptacle or a duplex receptacle for two appliances located within dedicated space for each appliance that, in normal use, is not easily moved from one place to another and that is cord-and-plug connected in accordance with 400.7(A)(6), (A)(7), or (A)(8).

Receptacles installed under the exceptions to 210.8(A)(5) shall not be considered as meeting the requirements of 210.52(G).

(3) Outdoors

Exception: Receptacles that are not readily accessible and are supplied by a dedicated branch circuit for electric snow-melting or deicing equipment shall be permitted to be installed in accordance with the applicable provisions of Article 426.

(4) Crawl spaces—at or below grade level
(5) Unfinished basements—for purposes of this section, unfinished basements are defined as portions or areas of the basement not intended as habitable rooms and limited to storage areas, work areas, and the like

Exception No. 1: Receptacles that are not readily accessible.

Exception No. 2: A single receptacle or a duplex receptacle for two appliances located within dedicated space for each appliance that, in normal use, is not easily moved from one place to another and that is cord-and-plug connected in accordance with 400.7(A)(6), (A)(7), or (A)(8).

Exception No. 3: A receptacle supplying only a permanently installed fire alarm system shall not be required to have ground-fault circuit-interrupter protection.

Receptacles installed under the exceptions to 210.8(A)(5) shall not be considered as meeting the requirements of 210.52(G).

(6) Kitchens—where the receptacles are installed to serve the countertop surfaces
(7) Wet bar sinks—where the receptacles are installed to serve the countertop surfaces and are located within 1.8 m (6 ft) of the outside edge of the wet bar sink.
(8) Boathouses

210.11 Branch Circuits Required. Branch circuits for lighting and for appliances, including motor-operated appliances, shall be provided to supply the loads computed in accordance with 220.3. In addition, branch circuits shall be provided for specific loads not covered by 220.3 where required elsewhere in this *Code* and for dwelling unit loads as specified in 210.11(C).

(A) Number of Branch Circuits. The minimum number of branch circuits shall be determined from the total computed load and the size or rating of the circuits used. In all installations, the number of circuits shall be sufficient to supply the load served. In no case shall the load on any circuit exceed the maximum specified by 220.4.

(B) Load Evenly Proportioned Among Branch Circuits. Where the load is computed on a volt-amperes/square meter or square foot basis, the wiring system up to and including the branch-circuit panelboard(s) shall be provided to serve not less than the calculated load. This load shall be evenly proportioned among multioutlet branch circuits within the panelboard(s). Branch-circuit overcurrent devices and circuits shall only be required to be installed to serve the connected load.

(C) Dwelling Units.

(1) Small-Appliance Branch Circuits. In addition to the number of branch circuits required by other parts of this section, two or more 20-ampere small-appliance branch circuits shall be provided for all receptacle outlets specified by 210.52(B).

(2) Laundry Branch Circuits. In addition to the number of branch circuits required by other parts of this section, at least one additional 20-ampere branch circuit shall be provided to supply the laundry receptacle outlet(s) required by 210.52(F). This circuit shall have no other outlets.

(3) Bathroom Branch Circuits. In addition to the number of branch circuits required by other parts of this section, at least one 20-ampere branch circuit shall be provided to supply the bathroom receptacle outlet(s). Such circuits shall have no other outlets.

Exception: Where the 20-ampere circuit supplies a single bathroom, outlets for other equipment within the same bathroom shall be permitted to be supplied in accordance with 210.23(A).

210.12 Arc-Fault Circuit-Interrupter Protection.

(A) Definition. An *arc-fault circuit interrupter* is a device intended to provide protection from the effects of arc faults by recognizing characteristics unique to arcing and by functioning to de-energize the circuit when an arc fault is detected.

(B) Dwelling Unit Bedrooms. All branch circuits that supply 125-volt, single-phase, 15- and 20-ampere outlets installed in dwelling unit bedrooms shall be protected by an arc-fault circuit interrupter listed to provide protection of the entire branch circuit.

II. BRANCH-CIRCUIT RATINGS

210.19 Conductors—Minimum Ampacity and Size.

(A) Branch Circuits Not More Than 600 Volts.

(1) General. Branch-circuit conductors shall have an ampacity not less than the maximum load to be served. Where a branch circuit supplies continuous loads or any combination of continuous and noncontinuous loads, the minimum branch-circuit conductor size, before the application of any adjustment or correction factors, shall have an allowable ampacity not less than the noncontinuous load plus 125 percent of the continuous load.

(2) Multioutlet Branch Circuits. Conductors of branch circuits supplying more than one receptacle for cord-and-plug-connected portable loads shall have an ampacity of not less than the rating of the branch circuit.

(3) Household Ranges and Cooking Appliances. Branch-circuit conductors supplying household ranges, wall-mounted ovens, counter-mounted cooking units, and other household cooking appliances shall have an ampacity not less than the rating of the branch circuit and not less than the maximum load to be served. For ranges of 8-3/4 kW or more rating, the minimum branch-circuit rating shall be 40 amperes.

Exception No. 1: Tap conductors supplying electric ranges, wall-mounted electric ovens, and counter-mounted electric cooking units from a 50-ampere branch circuit shall have an ampacity of not less than 20 and shall be sufficient for the load to be served. The taps shall not be longer than necessary for servicing the appliance.

Exception No. 2: The neutral conductor of a 3-wire branch circuit supplying a household electric range, a wall-mounted oven, or a counter-mounted cooking unit shall be permitted to be smaller than the ungrounded conductors where the maximum demand of a range of 8-3/4 kW or more rating has been computed according to Column C of Table 220.19, but shall have an ampacity of not less than 70 percent of the branch-circuit rating and shall not be smaller than 10 AWG.

(4) Other Loads. Branch-circuit conductors that supply loads other than those specified in 210.2 and other than cooking appliances as covered in 210.19(A)(3) shall have an ampacity sufficient for the loads served and shall not be smaller than 14 AWG.

210.21 Outlet Devices. Outlet devices shall have an ampere rating that is not less than the load to be served and shall comply with 210.21(A) and (B).

(B) Receptacles.

(1) Single Receptacle on an Individual Branch Circuit. A single receptacle installed on an individual branch circuit shall have an ampere rating not less than that of the branch circuit.

(2) Total Cord-and-Plug-Connected Load. Where connected to a branch circuit supplying two or more receptacles or outlets, a receptacle shall not supply a total cord-and-plug-connected load in excess of the maximum specified in Table 210.21(B)(2).

Table 210.21(B)(2) Maximum Cord-and-Plug-Connected Load to Receptacle

Circuit Rating (Amperes)	Receptacle Rating (Amperes)	Maximum Load (Amperes)
15 or 20	15	12
20	20	16
30	30	24

(3) Receptacle Ratings. Where connected to a branch circuit supplying two or more receptacles or outlets, receptacle ratings shall conform to the values listed in Table 210.21(B)(3), or where larger than 50 amperes, the receptacle rating shall not be less than the branch-circuit rating.

(4) Range Receptacle Rating. The ampere rating of a range receptacle shall be permitted to be based on a single range demand load as specified in Table 220.19.

210.23 Permissible Loads. In no case shall the load exceed the branch-circuit ampere rating. An individual branch circuit shall be permitted to supply any load for which it is rated. A branch circuit supplying two or more outlets or receptacles shall supply only the loads specified according to its size as specified in 210.23(A) through (D) and as summarized in 210.24 and Table 210.24.

(A) 15- and 20-Ampere Branch Circuits. A 15- or 20-ampere branch circuit shall be permitted to supply lighting units or other utilization equipment, or a combination of both, and shall comply with 210.23(A)(1) and (A)(2).

Exception: The small appliance branch circuits, laundry branch circuits, and bathroom branch circuits required in a dwelling unit(s) by 210.11(C)(1), (2), and (3) shall supply only the receptacle outlets specified in that section.

(1) Cord-and-Plug-Connected Equipment. The rating of any one cord-and-plug-connected utilization equipment shall not exceed 80 percent of the branch-circuit ampere rating.

(2) Utilization Equipment Fastened in Place. The total rating of utilization equipment fastened in place, other than

Table 210.21(B)(3) Receptacle Ratings for Various Size Circuits

Circuit Rating (Amperes)	Receptacle Rating (Amperes)
15	Not over 15
20	15 or 20
30	30
40	40 or 50
50	50

luminaires (lighting fixtures), shall not exceed 50 percent of the branch-circuit ampere rating where lighting units, cord-and-plug-connected utilization equipment not fastened in place, or both, are also supplied.

(B) 30-Ampere Branch Circuits. A 30-ampere branch circuit shall be permitted to supply fixed lighting units with heavy-duty lampholders in other than a dwelling unit(s) or utilization equipment in any occupancy. A rating of any one cord-and-plug-connected utilization equipment shall not exceed 80 percent of the branch-circuit ampere rating.

(C) 40- and 50-Ampere Branch Circuits. A 40- or 50-ampere branch circuit shall be permitted to supply cooking appliances that are fastened in place in any occupancy.

(D) Branch Circuits Larger Than 50 Amperes. Branch circuits larger than 50 amperes shall supply only nonlighting outlet loads.

210.24 Branch-Circuit Requirements—Summary. The requirements for circuits that have two or more outlets or receptacles, other than the receptacle circuits of 210.11(C)(1) and (2), are summarized in Table 210.24. This table provides only a summary of minimum requirements. See 210.19, 210.20, and 210.21 for the specific requirements applying to branch circuits.

210.25 Common Area Branch Circuits. Branch circuits in dwelling units shall supply only loads within that dwelling unit or loads associated only with that dwelling unit. Branch circuits required for the purpose of lighting, central alarm, signal, communications, or other needs for public or common areas of a two-family or multifamily dwelling shall not be supplied from equipment that supplies an individual dwelling unit.

III. REQUIRED OUTLETS

(See Chapter 17 for required receptacles, and Chapter 19 for required lighting outlets.)

Table 210.24 Summary of Branch-Circuit Requirements

Circuit Rating	15 A	20 A	30 A	40 A	50 A
Conductors (min. size):					
Circuit wires [1]	14	12	10	8	6
Taps	14	14	14	12	12
Fixture wires and cords—See 240.5					
Overcurrent Protection	15 A	20 A	30 A	40 A	50 A
Outlet devices:					
Lampholders permitted	Any type	Any type	Heavy duty	Heavy duty	Heavy duty
Receptacle rating [2]	15 max. A	15 or 20 A	30 A	40 or 50 A	50 A
Maximum Load	15 A	20 A	30 A	40 A	50 A
Permissible load	See 210.23(A)	See 210.23(A)	See 210.23(B)	See 210.23(C)	See 210.23(C)

[1]These gauges are for copper conductors.

[2]For receptacle rating of cord-connected electric-discharge luminaires (lighting fixtures), see 410.30(C).

CHAPTER 17

RECEPTACLE PROVISIONS

INTRODUCTION

This chapter contains provisions from Article 210—Branch Circuits and Article 406—Receptacles, Cord Connectors, and Attachment Plugs (Caps). The provisions from Article 210 covered in this chapter include 210.50 through 210.63. In those sections are specifications for installing receptacle outlets for both general and specific locations. General locations include: kitchens, family rooms, dining rooms, living rooms, bedrooms, and similar rooms and areas. Specific locations include: bathrooms, hallways, basements and garages, laundry areas, and kitchen counter-top areas.

Article 406 covers the rating, type, and installation of receptacles, cord connectors, and attachment plugs (cord caps). The requirements in Article 406 have been divided into two chapters of this pocket guide. This chapter covers all of the provisions except 406.3(D). Receptacle replacement provisions are located in Chapter 26.

ARTICLE 210
Branch Circuits

III. REQUIRED OUTLETS

210.50 General. Receptacle outlets shall be installed as specified in 210.52 through 210.63.

(A) Cord Pendants. A cord connector that is supplied by a permanently connected cord pendant shall be considered a receptacle outlet.

(B) Cord Connections. A receptacle outlet shall be installed wherever flexible cords with attachment plugs are used. Where flexible cords are permitted to be permanently connected, receptacles shall be permitted to be omitted for such cords.

(C) Appliance Outlets. Appliance receptacle outlets installed in a dwelling unit for specific appliances, such as laundry equipment, shall be installed within 1.8 m (6 ft) of the intended location of the appliance.

210.52 Dwelling Unit Receptacle Outlets. This section provides requirements for 125-volt, 15- and 20-ampere receptacle outlets. Receptacle outlets required by this section shall be in addition to any receptacle that is part of a luminaire (lighting fixture) or appliance, located within cabinets or cupboards, or located more than 1.7 m (5-1/2 ft) above the floor.

Permanently installed electric baseboard heaters equipped with factory-installed receptacle outlets or outlets provided as a separate assembly by the manufacturer shall be permitted as the required outlet or outlets for the wall space utilized by such permanently installed heaters. Such receptacle outlets shall not be connected to the heater circuits.

FPN: Listed baseboard heaters include instructions that may not permit their installation below receptacle outlets.

(A) General Provisions. In every kitchen, family room, dining room, living room, parlor, library, den, sunroom, bedroom, recreation room, or similar room or area of dwelling units, receptacle outlets shall be installed in accordance with the general provisions specified in 210.52(A)(1) through (A)(3).

(1) Spacing. Receptacles shall be installed so that no point measured horizontally along the floor line in any wall space is more than 1.8 m (6 ft) from a receptacle outlet.

(2) Wall Space. As used in this section, a wall space shall include the following:

(1) Any space 600 mm (2 ft) or more in width (including space measured around corners) and unbroken along the floor line by doorways, fireplaces, and similar openings
(2) The space occupied by fixed panels in exterior walls, excluding sliding panels
(3) The space afforded by fixed room dividers such as freestanding bar-type counters or railings

(3) Floor Receptacles. Receptacle outlets in floors shall not be counted as part of the required number of receptacle outlets unless located within 450 mm (18 in.) of the wall.

(B) Small Appliances.

(1) Receptacle Outlets Served. In the kitchen, pantry, breakfast room, dining room, or similar area of a dwelling unit, the two or more 20-ampere small-appliance branch circuits required by 210.11(C)(1) shall serve all receptacle outlets covered by 210.52(A) and (C) and receptacle outlets for refrigeration equipment.

Exception No. 1: In addition to the required receptacles specified by 210.52, switched receptacles supplied from a general-purpose branch circuit as defined in 210.70(A)(1), Exception No. 1, shall be permitted.

Exception No. 2: The receptacle outlet for refrigeration equipment shall be permitted to be supplied from an individual branch circuit rated 15 amperes or greater.

(2) No Other Outlets. The two or more small-appliance branch circuits specified in 210.52(B)(1) shall have no other outlets.

Exception No. 1: A receptacle installed solely for the electrical supply to and support of an electric clock in any of the rooms specified in 210.52(B)(1).

Exception No. 2: Receptacles installed to provide power for supplemental equipment and lighting on gas-fired ranges, ovens, or counter-mounted cooking units.

(3) Kitchen Receptacle Requirements. Receptacles installed in a kitchen to serve countertop surfaces shall be supplied by not fewer than two small-appliance branch circuits, either or both of which shall also be permitted to supply receptacle outlets in the same kitchen and in other rooms specified in 210.52(B)(1). Additional small-appliance branch circuits shall be permitted to supply receptacle outlets in the kitchen and other rooms specified in 210.52(B)(1). No small-appliance branch circuit shall serve more than one kitchen.

(C) Countertops. In kitchens and dining rooms of dwelling units, receptacle outlets for counter spaces shall be installed in accordance with 210.52(C)(1) through (5).

(1) Wall Counter Spaces. A receptacle outlet shall be installed at each wall counter space that is 300 mm (12 in.) or wider. Receptacle outlets shall be installed so that no point along the wall line is more than 600 mm (24 in.) measured horizontally from a receptacle outlet in that space.

(2) Island Counter Spaces. At least one receptacle outlet shall be installed at each island counter space with a long dimension of 600 mm (24 in.) or greater and a short dimension of 300 mm (12 in.) or greater.

(3) Peninsular Counter Spaces. At least one receptacle outlet shall be installed at each peninsular counter space with a long dimension of 600 mm (24 in.) or greater and a short dimension of 300 mm (12 in.) or greater. A peninsular countertop is measured from the connecting edge.

(4) Separate Spaces. Countertop spaces separated by range tops, refrigerators, or sinks shall be considered as separate countertop spaces in applying the requirements of 210.52(C)(1), (2), and (3).

(5) Receptacle Outlet Location. Receptacle outlets shall be located above, but not more than 500 mm (20 in.) above, the countertop. Receptacle outlets rendered not readily accessible by appliances fastened in place, appliance garages, or appliances occupying dedicated space shall not be considered as these required outlets.

Exception: To comply with the conditions specified in (a) or (b), receptacle outlets shall be permitted to be mounted not more than 300 mm (12 in.) below the countertop. Receptacles mounted below a countertop in accordance with this exception shall not be located where the countertop extends more than 150 mm (6 in.) beyond its support base.

(a) Construction for the physically impaired.
(b) On island and peninsular countertops where the countertop is flat across its entire surface (no backsplashes, dividers, etc.) and there are no means to mount a receptacle within 500 mm (20 in.) above the countertop, such as an overhead cabinet.

(D) Bathrooms. In dwelling units, at least one wall receptacle outlet shall be installed in bathrooms within 900 mm (3 ft) of the outside edge of each basin. The receptacle outlet shall be located on a wall or partition that is adjacent to the basin or basin countertop.

(E) Outdoor Outlets. For a one-family dwelling and each unit of a two-family dwelling that is at grade level, at least one receptacle outlet accessible at grade level and not more than 2.0 m (6-1/2 ft) above grade shall be installed at the front and back of the dwelling. See 210.8(A)(3).

(F) Laundry Areas. In dwelling units, at least one receptacle outlet shall be installed for the laundry.

Exception No. 1: In a dwelling unit that is an apartment or living area in a multifamily building where laundry facilities are provided on the premises and are available to all building occupants, a laundry receptacle shall not be required.

Exception No. 2: In other than one-family dwellings where laundry facilities are not to be installed or permitted, a laundry receptacle shall not be required.

(G) Basements and Garages. For a one-family dwelling, at least one receptacle outlet, in addition to any provided for laundry equipment, shall be installed in each basement and in each attached garage, and in each detached garage with electric power. See 210.8(A)(2) and (A)(5). Where a portion of the basement is finished into one or more habitable rooms, each separate unfinished portion shall have a receptacle outlet installed in accordance with this section.

(H) Hallways. In dwelling units, hallways of 3.0 m (10 ft) or more in length shall have at least one receptacle outlet.

As used in this subsection, the hall length shall be considered the length along the centerline of the hall without passing through a doorway.

210.63 Heating, Air-Conditioning, and Refrigeration Equipment Outlet. A 125-volt, single-phase, 15- or 20-ampere-rated receptacle outlet shall be installed at an accessible location for the servicing of heating, air-conditioning, and refrigeration equipment. The receptacle shall be located on the same level and within 7.5 m (25 ft) of the heating, air-conditioning, and refrigeration equipment. The receptacle outlet shall not be connected to the load side of the equipment disconnecting means.

ARTICLE 406

Receptacles, Cord Connectors, and Attachment Plugs (Caps)

406.1 Scope. This article covers the rating, type, and installation of receptacles, cord connectors, and attachment plugs (cord caps).

406.2 Receptacle Rating and Type.

(A) Receptacles. Receptacles shall be listed for the purpose and marked with the manufacturer's name or identification and voltage and ampere ratings.

(B) Rating. Receptacles and cord connectors shall be rated not less than 15 amperes, 125 volts, or 15 amperes, 250 volts, and shall be of a type not suitable for use as lampholders.

(C) Receptacles for Aluminum Conductors. Receptacles rated 20 amperes or less and designed for the direct connection of aluminum conductors shall be marked CO/ALR.

(D) Isolated Ground Receptacles. Receptacles incorporating an isolated grounding connection intended for the reduction of electrical noise (electromagnetic interference) as permitted in 250.146(D) shall be identified by an orange triangle located on the face of the receptacle.

(1) Receptacles so identified shall be used only with grounding conductors that are isolated in accordance with 250.146(D).

(2) Isolated ground receptacles installed in nonmetallic boxes shall be covered with a nonmetallic faceplate.

Exception: Where an isolated ground receptacle is installed in a nonmetallic box, a metal faceplate shall be permitted if the box contains a feature or accessory that permits the effective grounding of the faceplate.

406.3 General Installation Requirements. Receptacle outlets shall be located in branch circuits in accordance with Part III of Article 210. General installation requirements shall be in accordance with 406.3(A) through (F).

(A) Grounding Type. Receptacles installed on 15- and 20-ampere branch circuits shall be of the grounding type. Grounding-type receptacles shall be installed only on circuits of the voltage class and current for which they are rated, except as provided in Tables 210.21(B)(2) and (B)(3).

Exception: Nongrounding-type receptacles installed in accordance with 406.3(D).

(B) To Be Grounded. Receptacles and cord connectors that have grounding contacts shall have those contacts effectively grounded.

Exception No. 2: Replacement receptacles as permitted by 406.3(D).

(C) Methods of Grounding. The grounding contacts of receptacles and cord connectors shall be grounded by connection to the equipment grounding conductor of the circuit supplying the receptacle or cord connector.

The branch-circuit wiring method shall include or provide an equipment-grounding conductor to which the grounding contacts of the receptacle or cord connector shall be connected.

406.4 Receptacle Mounting. Receptacles shall be mounted in boxes or assemblies designed for the purpose, and such boxes or assemblies shall be securely fastened in place.

(A) Boxes That Are Set Back. Receptacles mounted in boxes that are set back of the wall surface, as permitted in 370.20, shall be installed so that the mounting yoke or strap of the receptacle is held rigidly at the surface of the wall.

(B) Boxes That Are Flush. Receptacles mounted in boxes that are flush with the wall surface or project therefrom shall be installed so that the mounting yoke or strap of the receptacle is held rigidly against the box or raised box cover.

(C) Receptacles Mounted on Covers. Receptacles mounted to and supported by a cover shall be held rigidly against the cover by more than one screw or shall be a device assembly or box cover listed and identified for securing by a single screw.

(D) Position of Receptacle Faces. After installation, receptacle faces shall be flush with or project from faceplates of insulating material and shall project a minimum of 0.4 mm (0.015 in.) from metal faceplates.

(E) Receptacles in Countertops and Similar Work Surfaces in Dwelling Units. Receptacles shall not be installed in a face-up position in countertops or similar work surfaces.

(F) Exposed Terminals. Receptacles shall be enclosed so that live wiring terminals are not exposed to contact.

406.5 Receptacle Faceplates (Cover Plates). Receptacle faceplates shall be installed so as to completely cover the opening and seat against the mounting surface.

(A) Metal faceplates shall be of ferrous metal not less than 0.76 mm (0.030 in.) in thickness or of nonferrous metal not less than 1.02 mm (0.040 in.) in thickness.

(B) Metal faceplates shall be grounded.

(C) Faceplates of insulating material shall be noncombustible and not less than 2.54 mm (0.010 in.) in thickness but shall be permitted to be less than 2.54 mm (0.010 in.) in thickness if formed or reinforced to provide adequate mechanical strength.

406.6 Attachment Plugs. All attachment plugs and cord connectors shall be listed for the purpose and marked with the manufacturer's name or identification and voltage and ampere ratings.

(A) Attachment plugs and cord connectors shall be constructed so that there are no exposed current-carrying parts except the prongs, blades, or pins. The cover for wire terminations shall be a part that is essential for the operation of an attachment plug or connector (dead-front construction).

(B) Attachment plugs shall be installed so that their prongs, blades, or pins are not energized unless inserted into an energized receptacle. No receptacle shall be installed so as to require an energized attachment plug as its source of supply.

406.8 Receptacles in Damp or Wet Locations.

(A) Damp Locations. A receptacle installed outdoors in a location protected from the weather or in other damp locations shall have an enclosure for the receptacle that is weatherproof when the receptacle is covered (attachment plug cap not inserted and receptacle covers closed).

An installation suitable for wet locations shall also be considered suitable for damp locations.

A receptacle shall be considered to be in a location protected from the weather where located under roofed open porches, canopies, marquees, and the like, and will not be subjected to a beating rain or water runoff.

(B) Wet Locations.

(1) 15- and 20-Ampere Outdoor Receptacles. 15- and 20-ampere, 125- and 250-volt receptacles installed outdoors in a wet location shall have an enclosure that is weatherproof whether or not the attachment plug cap is inserted.

(2) Other Receptacles. All other receptacles installed in a wet location shall comply with (a) or (b):

(a) A receptacle installed in a wet location where the product intended to be plugged into it is not attended while in use (e.g., sprinkler system controller, landscape lighting, holiday lights, and so forth) shall have an enclosure that is weatherproof with the attachment plug cap inserted or removed.

(b) A receptacle installed in a wet location where the product intended to be plugged into it will be attended while in use (e.g., portable tools, and so forth) shall have an enclosure that is weatherproof when the attachment plug is removed.

(C) Bathtub and Shower Space. A receptacle shall not be installed within a bathtub or shower space.

(E) Flush Mounting with Faceplate. The enclosure for a receptacle installed in an outlet box flush-mounted on a wall surface shall be made weatherproof by means of a weatherproof faceplate assembly that provides a watertight connection between the plate and the wall surface.

406.10 Connecting Receptacle Grounding Terminal to Box. The connection of the receptacle grounding terminal shall comply with 250.146.

CHAPTER 18

SWITCHES

INTRODUCTION

This chapter contains provisions from Article 404—Switches. Unlike the disconnect switches covered in Chapter 8, this chapter covers general-use snap switches. These switches are generally installed in device boxes or on box covers. Provisions from this article cover general-use snap switches (single-pole, three-way, four-way, double-pole, etc.), dimmers, and similar control switches. Specifications pertaining to the required locations for switches are in Article 210 (see Chapter 19).

ARTICLE 404
Switches

I. INSTALLATION

404.1 Scope. The provisions of this article shall apply to all switches, switching devices, and circuit breakers where used as switches.

404.2 Switch Connections.

(A) Three-Way and Four-Way Switches. Three-way and four-way switches shall be wired so that all switching is done only in the ungrounded circuit conductor. Where in metal raceways or metal-armored cables, wiring between switches and outlets shall be in accordance with 300.20(A).

Exception: Switch loops shall not require a grounded conductor.

(B) Grounded Conductors. Switches or circuit breakers shall not disconnect the grounded conductor of a circuit.

404.4 Wet Locations. A switch or circuit breaker in a wet location or outside of a building shall be enclosed in a weatherproof enclosure or cabinet that shall comply with 312.2(A). Switches shall not be installed within wet locations in tub or shower spaces unless installed as part of a listed tub or shower assembly.

404.9 Provisions for General-Use Snap Switches.

(A) Faceplates. Faceplates provided for snap switches mounted in boxes and other enclosures shall be installed so as to completely cover the opening and, where the switch is flush mounted, seat against the finished surface.

(B) Grounding. Snap switches, including dimmer and similar control switches, shall be effectively grounded and shall provide a means to ground metal faceplates, whether or not a metal faceplate is installed. Snap switches shall be considered effectively grounded if either of the following conditions is met.

(1) The switch is mounted with metal screws to a metal box or to a nonmetallic box with integral means for grounding devices.

(2) An equipment grounding conductor or equipment bonding jumper is connected to an equipment grounding termination of the snap switch.

Exception to (B): Where no grounding means exists within the snap-switch enclosure or where the wiring method does not include or provide an equipment ground, a snap switch without a grounding connection shall be permitted for replacement purposes only. A snap switch wired under the provisions of this exception and located within reach of earth, grade conducting floors, or other conducting surfaces shall be provided with a faceplate of nonconducting, noncombustible material.

404.10 Mounting of Snap Switches.

(A) Surface-Type. Snap switches used with open wiring on insulators shall be mounted on insulating material that separates the conductors at least 13 mm (1/2 in.) from the surface wired over.

(B) Box Mounted. Flush-type snap switches mounted in boxes that are set back of the wall surface as permitted in 314.20 shall be installed so that the extension plaster ears are seated against the surface of the wall. Flush-type snap switches mounted in boxes that are flush with the wall surface or project from it shall be installed so that the mounting yoke or strap of the switch is seated against the box.

404.14 Rating and Use of Snap Switches. Snap switches shall be used within their ratings and as indicated in 404.14(A) through (D).

FPN No. 1: For switches on signs and outline lighting, see 600.6.

FPN No. 2: For switches controlling motors, see 430.83, 430.109, and 430.110.

(A) Alternating Current General-Use Snap Switch. A form of general-use snap switch suitable only for use on ac circuits for controlling the following:

(1) Resistive and inductive loads, including electric-discharge lamps, not exceeding the ampere rating of the switch at the voltage involved

(2) Tungsten-filament lamp loads not exceeding the ampere rating of the switch at 120 volts

(3) Motor loads not exceeding 80 percent of the ampere rating of the switch at its rated voltage

(C) CO/ALR Snap Switches. Snap switches rated 20 amperes or less directly connected to aluminum conductors shall be listed and marked CO/ALR.

(E) Dimmer Switches. General-use dimmer switches shall be used only to control permanently installed incandescent luminaires (lighting fixtures) unless listed for the control of other loads and installed accordingly.

CHAPTER 19

LIGHTING REQUIREMENTS

INTRODUCTION

This chapter contains provisions from Article 210—Branch Circuits; Article 410—Luminaires (Lighting Fixtures), Lampholders, and Lamps; and Article 411—Lighting Systems Operating at 30 Volts or Less. This chapter covers the last section of Article 210, which provides requirements for lighting outlets and switches for dwelling units.

Article 410 covers luminaires (lighting fixtures), lampholders, pendants, incandescent filament lamps, and wiring and equipment forming part of such lamps, luminaires (fixtures), and lighting installations. Contained in this pocket guide are eight of the fifteen parts that are included in Article 410. The eight parts are: I General; II Luminaire (Fixture) Locations; III Provisions at Luminaire (Fixture) Outlet Boxes, Canopies, and Pans; IV Luminaire (Fixture) Supports; V Grounding; VI Wiring of Luminaires (Fixtures); XI Special Provisions for Flush and Recessed Luminaires (Fixtures); and XV Lighting Track.

Article 411 covers lighting systems operating at 30 volts or less and their associated components.

ARTICLE 210
Branch Circuits

III. REQUIRED OUTLETS

210.70 Lighting Outlets Required.

(A) Dwelling Units. In dwelling units, lighting outlets shall be installed in accordance with 210.70(A)(1), (2), and (3).

(1) Habitable Rooms. At least one wall switch-controlled lighting outlet shall be installed in every habitable room and bathroom.

Exception No. 1: In other than kitchens and bathrooms, one or more receptacles controlled by a wall switch shall be permitted in lieu of lighting outlets.

Exception No. 2: Lighting outlets shall be permitted to be controlled by occupancy sensors that are (1) in addition to wall switches or (2) located at a customary wall switch location and equipped with a manual override that will allow the sensor to function as a wall switch.

(2) Additional Locations. Additional lighting outlets shall be installed in accordance with (a), (b), and (c).

(a) At least one wall switch-controlled lighting outlet shall be installed in hallways, stairways, attached garages, and detached garages with electric power.

(b) For dwelling units, attached garages, and detached garages with electric power, at least one wall switch—controlled lighting outlet shall be installed to provide illumination on the exterior side of outdoor entrances or exits with grade level access. A vehicle door in a garage shall not be considered as an outdoor entrance or exit.

(c) Where one or more lighting outlet(s) are installed for interior stairways, there shall be a wall switch at each floor level, and landing level that includes an entry way, to control the lighting outlet(s) where the stairway between floor levels has six risers or more.

Exception to (a), (b), and (c): In hallways, stairways, and at outdoor entrances, remote, central, or automatic control of lighting shall be permitted.

(3) Storage or Equipment Spaces. For attics, underfloor spaces, utility rooms, and basements, at least one lighting outlet containing a switch or controlled by a wall switch shall be installed where these spaces are used for storage or contain equipment requiring servicing. At least one point of control shall be at the usual point of entry to these spaces. The lighting outlet shall be provided at or near the equipment requiring servicing.

ARTICLE 410
Luminaires (Lighting Fixtures), Lampholders, and Lamps

I. GENERAL

410.1 Scope. This article covers luminaires (lighting fixtures), lampholders, pendants, incandescent filament lamps, arc lamps, electric-discharge lamps, the wiring and equipment forming part of such lamps, luminaires (fixtures), and lighting installations.

II. LUMINAIRE (FIXTURE) LOCATIONS

410.4 Luminaires (Fixtures) in Specific Locations.

(A) Wet and Damp Locations. Luminaires (fixtures) installed in wet or damp locations shall be installed so that water cannot enter or accumulate in wiring compartments, lampholders, or other electrical parts. All luminaires (fixtures) installed in wet locations shall be marked, "Suitable for Wet Locations." All luminaires (fixtures) installed in damp locations shall be marked, "Suitable for Wet Locations" or "Suitable for Damp Locations."

(D) Bathtub and Shower Areas. No parts of cord-connected luminaires (fixtures), hanging luminaires (fixtures), lighting track, pendants, or ceiling-suspended (paddle) fans shall be located within a zone measured 900 mm (3 ft) horizontally and 2.5 m (8 ft) vertically from the top of the bathtub rim or shower stall threshold. This zone is all encompassing and includes the zone directly over the tub or shower stall.

410.5 Luminaires (Fixtures) Near Combustible Material.
Luminaires (fixtures) shall be constructed, installed, or
equipped with shades or guards so that combustible material
is not subjected to temperatures in excess of 90°C (194°F).

410.6 Luminaires (Fixtures) Over Combustible Material. Lampholders installed over highly combustible material shall be of the unswitched type. Unless an individual
switch is provided for each luminaire (fixture), lampholders
shall be located at least 2.5 m (8 ft) above the floor or shall
be located or guarded so that the lamps cannot be readily
removed or damaged.

410.8 Luminaires (Fixtures) in Clothes Closets.

(A) Definition.

Storage Space. The volume bounded by the sides and
back closet walls and planes extending from the closet
floor vertically to a height of 1.8 m (6 ft) or the highest
clothes-hanging rod and parallel to the walls at a horizontal
distance of 600 mm (24 in.) from the sides and back of the
closet walls, respectively, and continuing vertically to the
closet ceiling parallel to the walls at a horizontal distance of
300 mm (12 in.) or the width of the shelf, whichever is
greater; for a closet that permits access to both sides of a
hanging rod, this space includes the volume below the
highest rod extending 300 mm (12 in.) on either side of the
rod on a plane horizontal to the floor extending the entire
length of the rod.

   FPN: See Figure 410.8.

(B) Luminaire (Fixture) Types Permitted. Listed luminaires (fixtures) of the following types shall be permitted to
be installed in a closet:

(1) A surface-mounted or recessed incandescent luminaire (fixture) with a completely enclosed lamp
(2) A surface-mounted or recessed fluorescent luminaire
    (fixture)

(C) Luminaire (Fixture) Types Not Permitted. Incandescent luminaires (fixtures) with open or partially enclosed
lamps and pendant luminaires (fixtures) or lampholders
shall not be permitted.

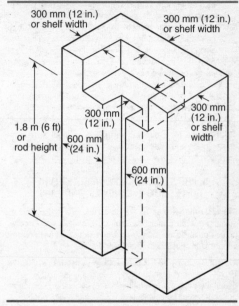

300 mm (12 in.) or shelf width

300 mm (12 in.) or shelf width

300 mm (12 in.) or shelf width

300 mm (12 in.)

1.8 m (6 ft) or rod height

600 mm (24 in.)

600 mm (24 in.)

Figure 410.8 Closet storage space.

(D) Location. Luminaires (fixtures) in clothes closets shall be permitted to be installed as follows:

(1) Surface-mounted incandescent luminaires (fixtures) installed on the wall above the door or on the ceiling, provided there is a minimum clearance of 300 mm (12 in.) between the luminaire (fixture) and the nearest point of a storage space

(2) Surface-mounted fluorescent luminaires (fixtures) installed on the wall above the door or on the ceiling, provided there is a minimum clearance of 150 mm (6 in.)

between the luminaire (fixture) and the nearest point of a storage space

(3) Recessed incandescent luminaires (fixtures) with a completely enclosed lamp installed in the wall or the ceiling, provided there is a minimum clearance of 150 mm (6 in.) between the luminaire (fixture) and the nearest point of a storage space

(4) Recessed fluorescent luminaires (fixtures) installed in the wall or the ceiling, provided there is a minimum clearance of 150 mm (6 in.) between the luminaire (fixture) and the nearest point of a storage space

410.9 Space for Cove Lighting. Coves shall have adequate space and shall be located so that lamps and equipment can be properly installed and maintained.

III. PROVISIONS AT LUMINAIRE (FIXTURE) OUTLET BOXES, CANOPIES, AND PANS

410.12 Outlet Boxes to Be Covered. In a completed installation, each outlet box shall be provided with a cover unless covered by means of a luminaire (fixture) canopy, lampholder, receptacle, or similar device.

410.13 Covering of Combustible Material at Outlet Boxes. Any combustible wall or ceiling finish exposed between the edge of a luminaire (fixture) canopy or pan and an outlet box shall be covered with noncombustible material.

410.14 Connection of Electric-Discharge Luminaires (Lighting Fixtures).

(A) Independent of the Outlet Box. Electric-discharge luminaires (lighting fixtures) supported independently of the outlet box shall be connected to the branch circuit through metal raceway, nonmetallic raceway, Type MC cable, Type AC cable, Type MI cable, nonmetallic sheathed cable, or by flexible cord as permitted in 410.30(B) or (C).

(B) Access to Boxes. Electric-discharge luminaires (lighting fixtures) surface mounted over concealed outlet, pull, or junction boxes shall be installed with suitable openings in back of the fixture to provide access to the boxes.

IV. LUMINAIRE (FIXTURE) SUPPORTS

410.15 Supports.

(A) General. Luminaires (fixtures) and lampholders shall be securely supported. A luminaire (fixture) that weighs more than 3 kg (6 lb) or exceeds 400 mm (16 in.) in any dimension shall not be supported by the screw shell of a lampholder.

(B) Metal Poles Supporting Luminaires (Lighting Fixtures). Metal poles shall be permitted to be used to support luminaires (lighting fixtures) and as a raceway to enclose supply conductors, provided the following conditions are met:

(1) A metal pole shall have a handhole not less than 50 mm × 100 mm (2 in. × 4 in.) with a raintight cover to provide access to the supply terminations within the pole or pole base.

Exception No. 1: No handhole shall be required in a pole 2.5 m (8 ft) or less in height above grade where the supply wiring method continues without splice or pull point, and where the interior of the pole and any splices are accessible by removing the luminaire (fixture).

Exception No. 2: No handhole shall be required in a metal pole 6.0 m (20 ft) or less in height above grade that is provided with a hinged base.

(2) Where raceway risers or cable is not installed within the pole, a threaded fitting or nipple shall be brazed or welded to the pole opposite the handhole for the supply connection.

(3) A metal pole shall be provided with a grounding terminal.

    a. A pole with a handhole shall have the grounding terminal accessible from the handhole.

    b. A pole with a hinged base shall have the grounding terminal accessible within the base.

Exception: No grounding terminal shall be required in a pole 2.5 m (8 ft) or less in height above grade where the supply wiring method continues without splice or pull, and where the interior of the pole and any splices are accessible by removing the luminaire (fixture).

(4) A pole with a hinged base shall have the hinged base and pole bonded together.

(5) Metal raceways or other equipment grounding conductors shall be bonded to the pole with an equipment grounding conductor recognized by 250.118 and sized in accordance with 250.122.

(6) Conductors in vertical metal poles used as raceway shall be supported as provided in 300.19.

410.16 Means of Support.

(A) Outlet Boxes. Outlet boxes or fittings installed as required by 314.23 shall be permitted to support luminaires (fixtures).

(B) Inspection. Luminaires (fixtures) shall be installed so that the connections between the luminaire (fixture) conductors and the circuit conductors can be inspected without requiring the disconnection of any part of the wiring unless the luminaires (fixtures) are connected by attachment plugs and receptacles.

(C) Suspended Ceilings. Framing members of suspended ceiling systems used to support luminaires (fixtures) shall be securely fastened to each other and shall be securely attached to the building structure at appropriate intervals. Luminaires (fixtures) shall be securely fastened to the ceiling framing member by mechanical means such as bolts, screws, or rivets. Listed clips identified for use with the type of ceiling framing member(s) and luminaire(s) [fixture(s)] shall also be permitted.

(D) Luminaire (Fixture) Studs. Luminaire (fixture) studs that are not a part of outlet boxes, hickeys, tripods, and crowfeet shall be made of steel, malleable iron, or other material suitable for the application.

(E) Insulating Joints. Insulating joints that are not designed to be mounted with screws or bolts shall have an exterior metal casing, insulated from both screw connections.

(F) Raceway Fittings. Raceway fittings used to support a luminaire(s) [lighting fixture(s)] shall be capable of supporting the weight of the complete fixture assembly and lamp(s).

(H) Trees. Outdoor luminaires (lighting fixtures) and associated equipment shall be permitted to be supported by trees.

> FPN No. 1: See 225.26 for restrictions for support of overhead conductors.

> FPN No. 2: See 300.5(D) for protection of conductors.

V. GROUNDING

410.17 General. Luminaires (fixtures) and lighting equipment shall be grounded as required in Article 250 and Part V of this article.

410.18 Exposed Luminaire (Fixture) Parts.

(A) Exposed Conductive Parts. Exposed metal parts shall be grounded or insulated from ground and other conducting surfaces or be inaccessible to unqualified personnel. Lamp tie wires, mounting screws, clips, and decorative bands on glass spaced at least 38 mm (1-1/2 in.) from lamp terminals shall not be required to be grounded.

(B) Made of Insulating Material. Luminaires (fixtures) directly wired or attached to outlets supplied by a wiring method that does not provide a ready means for grounding shall be made of insulating material and shall have no exposed conductive parts.

Exception: Replacement luminaires (fixtures) shall be permitted to connect an equipment grounding conductor from the outlet in compliance with 250.130(C). The luminaire (fixture) shall then be grounded in accordance with 410.18(A).

410.21 Methods of Grounding. Luminaires (fixtures) and equipment shall be considered grounded where mechanically connected to an equipment grounding conductor as specified in 250.118 and sized in accordance with 250.122.

VI. WIRING OF LUMINAIRES (FIXTURES)

410.23 Polarization of Luminaires (Fixtures). Luminaires (fixtures) shall be wired so that the screw shells of lampholders are connected to the same luminaire (fixture) or circuit conductor or terminal. The grounded conductor, where connected to a screw-shell lampholder, shall be connected to the screw shell.

410.28 Protection of Conductors and Insulation.

(A) Properly Secured. Conductors shall be secured in a manner that does not tend to cut or abrade the insulation.

(B) Protection Through Metal. Conductor insulation shall be protected from abrasion where it passes through metal.

(C) Luminaire (Fixture) Stems. Splices and taps shall not be located within luminaire (fixture) arms or stems.

(D) Splices and Taps. No unnecessary splices or taps shall be made within or on a luminaire (fixture).

410.31 Luminaires (Fixtures) as Raceways. Luminaires (fixtures) shall not be used as a raceway for circuit conductors unless listed and marked for use as a raceway.

410.33 Branch Circuit Conductors and Ballasts. Branch-circuit conductors within 75 mm (3 in.) of a ballast shall have an insulation temperature rating not lower than 90°C (194°F) unless supplying a luminaire (fixture) listed and marked as suitable for a different insulation temperature.

XI. SPECIAL PROVISIONS FOR FLUSH AND RECESSED LUMINAIRES (FIXTURES)

410.64 General. Luminaires (fixtures) installed in recessed cavities in walls or ceilings shall comply with 410.65 through 410.72.

410.65 Temperature.

(A) Combustible Material. Luminaires (fixtures) shall be installed so that adjacent combustible material will not be subjected to temperatures in excess of 90°C (194°F).

(B) Fire-Resistant Construction. Where a luminaire (fixture) is recessed in fire-resistant material in a building of fire-resistant construction, a temperature higher than 90°C (194°F) but not higher than 150°C (302°F) shall be considered acceptable if the luminaire (fixture) is plainly marked that it is listed for that service.

(C) Recessed Incandescent Luminaires (Fixtures). Incandescent luminaires (fixtures) shall have thermal protection and shall be identified as thermally protected.

Exception No. 1: Thermal protection shall not be required in a recessed luminaire (fixture) identified for use and installed in poured concrete.

Exception No. 2: Thermal protection shall not be required in a recessed luminaire (fixture) whose design, construction, and thermal performance characteristics are equivalent to a thermally protected luminaire (fixture) and are identified as inherently protected.

410.66 Clearance and Installation.

(A) Clearance.

(1) Non-Type IC. A recessed luminaire (fixture) that is not identified for contact with insulation shall have all recessed parts spaced not less than 13 mm (1/2 in.) from combustible materials. The points of support and the trim finishing off the opening in the ceiling or wall surface shall be permitted to be in contact with combustible materials.

(2) Type IC. A recessed luminaire (fixture) that is identified for contact with insulation, Type IC, shall be permitted to be in contact with combustible materials at recessed parts, points of support, and portions passing through or finishing off the opening in the building structure.

(B) Installation. Thermal insulation shall not be installed above a recessed luminaire (fixture) or within 75 mm (3 in.) of the recessed luminaire's (fixture's) enclosure, wiring compartment, or ballast unless it is identified for contact with insulation, Type IC.

XV. LIGHTING TRACK

410.100 Definition.

Lighting Track. A manufactured assembly designed to support and energize luminaires (lighting fixtures) that are capable of being readily repositioned on the track. Its length may be altered by the addition or subtraction of sections of track.

410.101 Installation.

(A) Lighting Track. Lighting track shall be permanently installed and permanently connected to a branch circuit. Only lighting track fittings shall be installed on lighting track. Lighting track fittings shall not be equipped with general-purpose receptacles.

(B) Connected Load. The connected load on lighting track shall not exceed the rating of the track. Lighting track shall be supplied by a branch circuit having a rating not more than that of the track.

(C) Locations Not Permitted. Lighting track shall not be installed in the following locations:

(1) Where likely to be subjected to physical damage
(2) In wet or damp locations
(6) Where concealed
(7) Where extended through walls or partitions
(8) Less than 1.5 m (5 ft) above the finished floor except where protected from physical damage or track operating at less than 30 volts rms open-circuit voltage
(9) Within the zone measured 900 mm (3 ft) horizontally and 2.5 m (8 ft) vertically from the top of the bathtub rim

(D) Support. Fittings identified for use on lighting track shall be designed specifically for the track on which they are to be installed. They shall be securely fastened to the track, shall maintain polarization and grounding, and shall be designed to be suspended directly from the track.

410.104 Fastening. Lighting track shall be securely mounted so that each fastening will be suitable for supporting the maximum weight of luminaires (fixtures) that can be installed. Unless identified for supports at greater intervals, a

single section 1.2 m (4 ft) or shorter in length shall have two supports, and, where installed in a continuous row, each individual section of not more than 1.2 m (4 ft) in length shall have one additional support.

410.105 Construction Requirements.

(B) Grounding. Lighting track shall be grounded in accordance with Article 250, and the track sections shall be securely coupled to maintain continuity of the circuitry, polarization, and grounding throughout.

ARTICLE 411
Lighting Systems Operating at 30 Volts or Less

411.1 Scope. This article covers lighting systems operating at 30 volts or less and their associated components.

411.2 Definition.

Lighting Systems Operating at 30 Volts or Less. A lighting system consisting of an isolating power supply operating at 30 volts (42.4 volts peak) or less, under any load condition, with one or more secondary circuits, each limited to 25 amperes maximum, supplying luminaires (lighting fixtures) and associated equipment identified for the use.

411.3 Listing Required. Lighting systems operating at 30 volts or less shall be listed for the purpose.

411.4 Locations Not Permitted. Lighting systems operating at 30 volts or less shall not be installed (1) where concealed or extended through a building wall, unless using a wiring method specified in Chapter 3, or (2) within 3.0 m (10 ft) of pools, spas, fountains, or similar locations, except as permitted by Article 680.

411.5 Secondary Circuits.

(A) Grounding. Secondary circuits shall not be grounded.

(B) Isolation. The secondary circuit shall be insulated from the branch circuit by an isolating transformer.

(C) Bare Conductors. Exposed bare conductors and current-carrying parts shall be permitted. Bare conductors shall not be installed less than 2.1 m (7 ft) above the finished floor, unless specifically listed for a lower installation height.

411.6 Branch Circuit. Lighting systems operating at 30 volts or less shall be supplied from a maximum 20-ampere branch circuit.

CHAPTER 20

APPLIANCES

INTRODUCTION

This chapter contains provisions from Article 422—Appliances. Article 422 covers electric appliances used in any occupancy. Installation provisions cover branch-circuit ratings and overcurrent protection. Flexible-cord requirements are provided for electrically operated kitchen waste disposers; built-in dishwashers and trash compactors; and wall-mounted ovens and counter-mounted cooking units. Appliance disconnecting specifications are covered in Part III of Article 422.

ARTICLE 422
Appliances

I. GENERAL

422.1 Scope. This article covers electric appliances used in any occupancy.

II. INSTALLATION

422.10 Branch-Circuit Rating. This section specifies the ratings of branch circuits capable of carrying appliance current without overheating under the conditions specified.

(A) Individual Circuits. The rating of an individual branch circuit shall not be less than the marked rating of the appliance or the marked rating of an appliance having combined loads as provided in 422.62.

The rating of an individual branch circuit for motor-operated appliances not having a marked rating shall be in accordance with Part II of Article 430.

The branch-circuit rating for an appliance that is continuously loaded, other than a motor-operated appliance, shall not be less than 125 percent of the marked rating, or not less than 100 percent of the marked rating if the branch-circuit device and its assembly are listed for continuous loading at 100 percent of its rating.

Branch circuits for household cooking appliances shall be permitted to be in accordance with Table 220.19.

(B) Circuits Supplying Two or More Loads. For branch circuits supplying appliance and other loads, the rating shall be determined in accordance with 210.23.

422.11 Overcurrent Protection. Appliances shall be protected against overcurrent in accordance with 422.11(A) through (G) and 422.10.

(A) Branch-Circuit Overcurrent Protection. Branch circuits shall be protected in accordance with 240.4.

If a protective device rating is marked on an appliance, the branch-circuit overcurrent device rating shall not exceed the protective device rating marked on the appliance.

(B) Household-Type Appliance with Surface Heating Elements. A household-type appliance with surface heating elements having a maximum demand of more than 60 amperes computed in accordance with Table 220.19 shall have its power supply subdivided into two or more circuits, each of which shall be provided with overcurrent protection rated at not over 50 amperes.

(E) Single Nonmotor-Operated Appliance. If the branch circuit supplies a single non–motor-operated appliance, the rating of overcurrent protection shall

(1) Not exceed that marked on the appliance;
(2) If the overcurrent protection rating is not marked and the appliance is rated 13.3 amperes or less, not exceed 20 amperes; or
(3) If the overcurrent protection rating is not marked and the appliance is rated over 13.3 amperes, not exceed 150 percent of the appliance rated current. Where 150 percent of the appliance rating does not correspond to a standard overcurrent device ampere rating, the next higher standard rating shall be permitted.

(G) Motor-Operated Appliances. Motors of motor-operated appliances shall be provided with overload protection in accordance with Part III of Article 430. Hermetic refrigerant motor-compressors in air-conditioning or refrigerating equipment shall be provided with overload protection in accordance with Part VI of Article 440. Where appliance overcurrent protective devices that are separate from the appliance are required, data for selection of these devices shall be marked on the appliance. The minimum marking shall be that specified in 430.7 and 440.4.

422.12 Central Heating Equipment. Central heating equipment other than fixed electric space-heating equipment shall be supplied by an individual branch circuit.

Exception: Auxiliary equipment, such as a pump, valve, humidifier, or electrostatic air cleaner directly associated with the heating equipment, shall be permitted to be connected to the same branch circuit.

422.13 Storage-Type Water Heaters. A branch circuit sup-plying a fixed storage-type water heater that has a capacity of 450 L (120 gal) or less shall have a rating not less than 125 percent of the nameplate rating of the water heater.

422.15 Central Vacuum Outlet Assemblies.

(A) Listed central vacuum outlet assemblies shall be per-mitted to be connected to a branch circuit in accordance with 210.23(A).

(B) The ampacity of the connecting conductors shall not be less than the ampacity of the branch circuit conductors to which they are connected.

(C) An equipment grounding conductor shall be used where the central vacuum outlet assembly has accessible non–current-carrying metal parts.

422.16 Flexible Cords.

(A) General. Flexible cord shall be permitted (1) for the connection of appliances to facilitate their frequent inter-change or to prevent the transmission of noise or vibration or (2) to facilitate the removal or disconnection of appli-ances that are fastened in place, where the fastening means and mechanical connections are specifically de-signed to permit ready removal for maintenance or repair and the appliance is intended or identified for flexible cord connection.

(B) Specific Appliances.

(1) Electrically Operated Kitchen Waste Disposers. Electrically operated kitchen waste disposers shall be per-mitted to be cord-and-plug connected with a flexible cord identified as suitable for the purpose in the installation in-structions of the appliance manufacturer, where all of the following conditions are met.

(1) The flexible cord shall be terminated with a grounding type attachment plug.

Exception: A listed kitchen waste disposer distinctly marked to identify it as protected by a system of double in-

sulation, or its equivalent, shall not be required to be terminated with a grounding-type attachment plug.

(2) The length of the cord shall not be less than 450 mm (18 in.) and not over 900 mm (36 in.).

(3) Receptacles shall be located to avoid physical damage to the flexible cord.

(4) The receptacle shall be accessible.

(2) Built-in Dishwashers and Trash Compactors. Built-in dishwashers and trash compactors shall be permitted to be cord-and-plug connected with a flexible cord identified as suitable for the purpose in the installation instructions of the appliance manufacturer where all of the following conditions are met.

(1) The flexible cord shall be terminated with a grounding-type attachment plug.

Exception: A listed dishwasher or trash compactor distinctly marked to identify it as protected by a system of double insulation, or its equivalent, shall not be required to be terminated with a grounding-type attachment plug.

(2) The length of the cord shall be 0.9 m to 1.2 m (3 ft to 4 ft) measured from the face of the attachment plug to the plane of the rear of the appliance.

(3) Receptacles shall be located to avoid physical damage to the flexible cord.

(4) The receptacle shall be located in the space occupied by the appliance or adjacent thereto.

(5) The receptacle shall be accessible.

(3) Wall-Mounted Ovens and Counter-Mounted Cooking Units. Wall-mounted ovens and counter-mounted cooking units complete with provisions for mounting and for making electrical connections shall be permitted to be permanently connected or, only for ease in servicing or for installation, cord-and-plug connected.

A separable connector or a plug and receptacle combination in the supply line to an oven or cooking unit shall be approved for the temperature of the space in which it is located.

422.18 Support of Ceiling-Suspended (Paddle) Fans.

(A) Ceiling-Suspended (Paddle) Fans 16 kg (35 lb) or Less. Ceiling-suspended (paddle) fans that do not exceed 16 kg (35 lb) in weight, with or without accessories, shall be permitted to be supported by outlet boxes identified for such use and supported in accordance with 314.23 and 314.27.

(B) Ceiling-Suspended (Paddle) Fans Exceeding 16 kg (35 lb). Ceiling-suspended (paddle) fans exceeding 16 kg (35 lb) in weight, with or without accessories, shall be supported independently of the outlet box. See 314.23.

Exception: Listed outlet boxes or outlet box systems that are identified for the purpose shall be permitted to support ceiling-suspended fans, with or without accessories, that weigh 32 kg (70 lb) or less.

422.20 Other Installation Methods. Appliances employing methods of installation other than covered by this article shall be permitted to be used only by special permission.

III. DISCONNECTING MEANS

422.30 General. A means shall be provided to disconnect each appliance from all ungrounded conductors in accordance with the following sections of Part III. If an appliance is supplied by more than one source, the disconnecting means shall be grouped and identified.

422.31 Disconnection of Permanently Connected Appliances.

(A) Rated at Not Over 300 Volt-Amperes or 1/8 Horsepower. For permanently connected appliances rated at not over 300 volt-amperes or 1/8 hp, the branch-circuit overcurrent device shall be permitted to serve as the disconnecting means.

(B) Appliances Rated Over 300 Volt-Amperes or 1/8 Horsepower. For permanently connected appliances rated over 300 volt-amperes or 1/8 hp, the branch-circuit switch or circuit breaker shall be permitted to serve as the disconnecting means where the switch or circuit breaker is within sight from the appliance or is capable of being locked in the open position.

422.32 Disconnecting Means for Motor-Driven Appliance. If a switch or circuit breaker serves as the disconnecting means for a permanently connected motor-driven appliance of more than 1/8 hp, it shall be located within sight from the motor controller and shall comply with Part X of Article 430.

Exception: If a motor-driven appliance of more than 1/8 hp is provided with a unit switch that complies with 422.34(A), (B), (C), or (D), the switch or circuit breaker serving as the other disconnecting means shall be permitted to be out of sight from the motor controller.

422.33 Disconnection of Cord-and-Plug-Connected Appliances.

(A) Separable Connector or an Attachment Plug and Receptacle. For cord-and-plug-connected appliances, an accessible separable connector or an accessible plug and receptacle shall be permitted to serve as the disconnecting means. Where the separable connector or plug and receptacle are not accessible, cord-and-plug-connected appliances shall be provided with disconnecting means in accordance with 422.31.

(B) Connection at the Rear Base of a Range. For cord-and-plug-connected household electric ranges, an attachment plug and receptacle connection at the rear base of a range, if it is accessible from the front by removal of a drawer, shall be considered as meeting the intent of 422.33(A).

(C) Rating. The rating of a receptacle or of a separable connector shall not be less than the rating of any appliance connected thereto.

422.34 Unit Switch(es) as Disconnecting Means. A unit switch(es) with a marked-off position that is a part of an appliance and disconnects all ungrounded conductors shall be permitted as the disconnecting means required by this article where other means for disconnection are provided in the following types of occupancies.

(A) Multifamily Dwellings. In multifamily dwellings, the other disconnecting means shall be within the dwelling unit, or on the same floor as the dwelling unit in which the appliance is installed, and shall be permitted to control lamps and other appliances.

(B) Two-Family Dwellings. In two-family dwellings, the other disconnecting means shall be permitted either inside or outside of the dwelling unit in which the appliance is installed. In this case, an individual switch or circuit breaker for the dwelling unit shall be permitted and shall also be permitted to control lamps and other appliances.

(C) One-Family Dwellings. In one-family dwellings, the service disconnecting means shall be permitted to be the other disconnecting means.

422.35 Switch and Circuit Breaker to Be Indicating. Switches and circuit breakers used as disconnecting means shall be of the indicating type.

CHAPTER 21

ELECTRIC SPACE-HEATING EQUIPMENT

INTRODUCTION

This chapter contains provisions from Article 424—Fixed Electric Space-Heating Equipment. Article 424 covers fixed electric equipment used for space heating. Heating equipment covered in this article includes: heating cable, unit heaters, central systems, and other approved fixed electric space-heating equipment. Contained in this pocket guide are four of the nine parts that are included in Article 424. The four parts are: I General; II Installation; III Control and Protection of Fixed Electric Space-Heating Equipment; and V Electric Space-Heating Cables. Room air-conditioning equipment is covered in Article 440 (see Chapter 22).

ARTICLE 424
Fixed Electric Space-Heating Equipment

I. GENERAL

424.1 Scope. This article covers fixed electric equipment used for space heating. For the purpose of this article, heating equipment shall include heating cable, unit heaters, boilers, central systems, or other approved fixed electric space-heating equipment. This article shall not apply to process heating and room air conditioning.

424.3 Branch Circuits.

(A) Branch-Circuit Requirements. Individual branch circuits shall be permitted to supply any size fixed electric space-heating equipment.

Branch circuits supplying two or more outlets for fixed electric space-heating equipment shall be rated 15, 20, 25, or 30 amperes. In other than residential occupancies, fixed infrared heating equipment shall be permitted to be supplied from branch circuits rated not over 50 amperes.

(B) Branch-Circuit Sizing. The ampacity of the branch-circuit conductors and the rating or setting of overcurrent protective devices supplying fixed electric space-heating equipment consisting of resistance elements with or without a motor shall not be less than 125 percent of the total load of the motors and the heaters. The rating or setting of overcurrent protective devices shall be permitted in accordance with 240.4(B). A contactor, thermostat, relay, or similar device, listed for continuous operation at 100 percent of its rating, shall be permitted to supply its full-rated load as provided in 210.19(A), Exception.

The size of the branch-circuit conductors and overcurrent protective devices supplying fixed electric space-heating equipment, including a hermetic refrigerant motor-compressor with or without resistance units, shall becomputed in accordance with 440.34 and 440.35. The provisions of this section shall not apply to conductors that form an integral part of approved fixed electric space-heating equipment.

II. INSTALLATION

424.9 General. All fixed electric space-heating equipment shall be installed in an approved manner.

Permanently installed electric baseboard heaters equipped with factory-installed receptacle outlets, or outlets provided as a separate listed assembly, shall be permitted in lieu of a receptacle outlet(s) that is required by 210.50(B). Such receptacle outlets shall not be connected to the heater circuits.

> FPN: Listed baseboard heaters include instructions that may not permit their installation below receptacle outlets.

424.12 Locations.

(A) Exposed to Physical Damage. Where subject to physical damage, fixed electric space-heating equipment shall be protected in an approved manner.

(B) Damp or Wet Locations. Heaters and related equipment installed in damp or wet locations shall be approved for such locations and shall be constructed and installed so that water or other liquids cannot enter or accumulate in or on wired sections, electrical components, or ductwork.

> FPN No. 1: See 110.11 for equipment exposed to deteriorating agents.

> FPN No. 2: See 680.27(C) for pool deck areas.

424.13 Spacing from Combustible Materials. Fixed electric space-heating equipment shall be installed to provide the required spacing between the equipment and adjacent combustible material, unless it has been found to be acceptable where installed in direct contact with combustible material.

III. CONTROL AND PROTECTION OF FIXED ELECTRIC SPACE-HEATING EQUIPMENT

424.19 Disconnecting Means. Means shall be provided to disconnect the heater, motor controller(s), and supplementary overcurrent protective device(s) of all fixed electric space-heating equipment from all ungrounded conductors. Where heating equipment is supplied by more than one source, the disconnecting means shall be grouped and marked.

(A) Heating Equipment with Supplementary Overcurrent Protection. The disconnecting means for fixed electric space-heating equipment with supplementary overcurrent protection shall be within sight from the supplementary overcurrent protective device(s), on the supply side of these devices, if fuses, and, in addition, shall comply with either 424.19(A)(1) or (2).

(1) Heater Containing No Motor Rated Over 1/8 Horsepower. The above disconnecting means or unit switches complying with 424.19(C) shall be permitted to serve as the required disconnecting means for both the motor controller(s) and heater under either item (1) or (2):

(1) The disconnecting means provided is also within sight from the motor controller(s) and the heater.
(2) The disconnecting means provided shall be capable of being locked in the open position.

(2) Heater Containing a Motor(s) Rated Over 1/8 Horsepower. The above disconnecting means shall be permitted to serve as the required disconnecting means for both the motor controller(s) and heater by one of the means specified in items (1) through (4):

(1) Where the disconnecting means is also in sight from the motor controller(s) and the heater.
(2) Where the disconnecting means is not within sight from the heater, a separate disconnecting means shall be installed, or the disconnecting means shall be capable of being locked in the open position, or unit switches complying with 424.19(C) shall be permitted.
(3) Where the disconnecting means is not within sight from the motor controller location, a disconnecting means complying with 430.102 shall be provided.

(4) Where the motor is not in sight from the motor controller location, 430.102(B) shall apply.

(B) Heating Equipment Without Supplementary Overcurrent Protection.

(1) Without Motor or with Motor Not Over 1/8 Horsepower. For fixed electric space-heating equipment without a motor rated over 1/8 hp, the branch-circuit switch or circuit breaker shall be permitted to serve as the disconnecting means where the switch or circuit breaker is within sight from the heater or is capable of being locked in the open position.

(2) Over 1/8 Horsepower. For motor-driven electric space-heating equipment with a motor rated over 1/8 hp, a disconnecting means shall be located within sight from the motor controller or shall be permitted to comply with the requirements in 424.19(A)(2).

(C) Unit Switch(es) as Disconnecting Means. A unit switch(es) with a marked "off" position that is part of a fixed heater and disconnects all ungrounded conductors shall be permitted as the disconnecting means required by this article where other means for disconnection are provided in the types of occupancies in 424.19(C)(1) through (C)(4).

(1) Multifamily Dwellings. In multifamily dwellings, the other disconnecting means shall be within the dwelling unit, or on the same floor as the dwelling unit in which the fixed heater is installed, and shall also be permitted to control lamps and appliances.

(2) Two-Family Dwellings. In two-family dwellings, the other disconnecting means shall be permitted either inside or outside of the dwelling unit in which the fixed heater is installed. In this case, an individual switch or circuit breaker for the dwelling unit shall be permitted and shall also be permitted to control lamps and appliances.

(3) One-Family Dwellings. In one-family dwellings, the service disconnecting means shall be permitted to be the other disconnecting means.

424.20 Thermostatically Controlled Switching Devices.

(A) Serving as Both Controllers and Disconnecting Means. Thermostatically controlled switching devices and combination thermostats and manually controlled switches shall be permitted to serve as both controllers and disconnecting means, provided all of the following conditions are met:

(1) Provided with a marked "off" position
(2) Directly open all ungrounded conductors when manually placed in the "off" position
(3) Designed so that the circuit cannot be energized automatically after the device has been manually placed in the "off" position
(4) Located as specified in 424.19

(B) Thermostats That Do Not Directly Interrupt All Ungrounded Conductor. Thermostats that do not directly interrupt all ungrounded conductors and thermostats that operate remote-control circuits shall not be required to meet the requirements of 424.20(A). These devices shall not be permitted as the disconnecting means.

424.21 Switch and Circuit Breaker to Be Indicating. Switches and circuit breakers used as disconnecting means shall be of the indicating type.

424.22 Overcurrent Protection.

(A) Branch-Circuit Devices. Electric space-heating equipment, other than such motor-operated equipment as required by Articles 430 and 440 to have additional overcurrent protection, shall be permitted to be protected against overcurrent where supplied by one of the branch circuits in Article 210.

V. ELECTRIC SPACE-HEATING CABLES

424.35 Marking of Heating Cables. Each unit shall be marked with the identifying name or identification symbol, catalog number, and ratings in volts and watts or in volts and amperes.

Each unit length of heating cable shall have a permanent legible marking on each nonheating lead located within 75 mm (3 in.) of the terminal end. The lead wire shall have the following color identification to indicate the circuit voltage on which it is to be used:

(1) 120 volt, nominal—yellow
(2) 208 volt, nominal—blue
(3) 240 volt, nominal—red
(4) 277 volt, nominal—brown
(5) 480 volt, nominal—orange

424.36 Clearances of Wiring in Ceilings. Wiring located above heated ceilings shall be spaced not less than 50 mm (2 in.) above the heated ceiling and shall be considered as operating at an ambient temperature of 50°C (122°F). The ampacity of conductors shall be computed on the basis of the correction factors shown in the 0–2000 volt ampacity tables of Article 310. If this wiring is located above thermal insulation having a minimum thickness of 50 mm (2 in.), the wiring shall not require correction for temperature.

424.37 Location of Branch-Circuit and Feeder Wiring in Exterior Walls. Wiring methods shall comply with Article 300 and 310.10.

424.38 Area Restrictions.

(A) Shall Not Extend Beyond the Room or Area. Heating cables shall not extend beyond the room or area in which they originate.

(B) Uses Prohibited. Heating cables shall not be installed in the following:

(1) In closets
(2) Over walls
(3) Over partitions that extend to the ceiling, unless they are isolated single runs of embedded cable
(4) Over cabinets whose clearance from the ceiling is less than the minimum horizontal dimension of the cabinet to the nearest cabinet edge that is open to the room or area

(C) In Closet Ceilings as Low-Temperature Heat Sources to Control Relative Humidity. The provisions of 424.38(B) shall not prevent the use of cable in closet ceilings as low-temperature heat sources to control relative humidity,

242

provided they are used only in those portions of the ceiling that are unobstructed to the floor by shelves or other permanent luminaires (fixtures).

424.39 Clearance from Other Objects and Openings. Heating elements of cables shall be separated at least 200 mm (8 in.) from the edge of outlet boxes and junction boxes that are to be used for mounting surface luminaires (lighting fixtures). A clearance of not less than 50 mm (2 in.) shall be provided from recessed luminaires (fixtures) and their trims, ventilating openings, and other such openings in room surfaces. Sufficient area shall be provided to ensure that no heating cable will be covered by any surface-mounted units.

424.40 Splices. Embedded cables shall be spliced only where necessary and only by approved means, and in no case shall the length of the heating cable be altered.

424.41 Installation of Heating Cables on Dry Board, in Plaster, and on Concrete Ceilings.

(A) In Walls. Cables shall not be installed in walls unless it is necessary for an isolated single run of cable to be installed down a vertical surface to reach a dropped ceiling.

(B) Adjacent Runs. Adjacent runs of cable not exceeding 9 watts/m (2-3/4 watts/ft) shall not be installed less than 38 mm (1-1/2 in.) on centers.

(C) Surfaces to Be Applied. Heating cables shall be applied only to gypsum board, plaster lath, or other fire-resistant material. With metal lath or other electrically conductive surfaces, a coat of plaster shall be applied to completely separate the metal lath or conductive surface from the cable.

(D) Splices. All heating cables, the splice between the heating cable and nonheating leads, and 75-mm (3-in.) minimum of the nonheating lead at the splice shall be embedded in plaster or dry board in the same manner as the heating cable.

(E) Ceiling Surface. The entire ceiling surface shall have a finish of thermally noninsulating sand plaster that has a nominal thickness of 13 mm (1/2 in.), or other noninsulating material identified as suitable for this use and applied according to specified thickness and directions.

(F) Secured. Cables shall be secured by means of approved stapling, tape, plaster, nonmetallic spreaders, or other approved means either at intervals not exceeding 400 mm (16 in.) or at intervals not exceeding 1.8 m (6 ft) for cables identified for such use. Staples or metal fasteners that straddle the cable shall not be used with metal lath or other electrically conductive surfaces.

(G) Dry Board Installations. In dry board installations, the entire ceiling below the heating cable shall be covered with gypsum board not exceeding 13 mm (1/2 in.) thickness. The void between the upper layer of gypsum board, plaster lath, or other fire-resistant material and the surface layer of gypsum board shall be completely filled with thermally conductive, nonshrinking plaster or other approved material or equivalent thermal conductivity.

(H) Free from Contact with Conductive Surfaces. Cables shall be kept free from contact with metal or other electrically conductive surfaces.

(I) Joists. In dry board applications, cable shall be installed parallel to the joist, leaving a clear space centered under the joist of 65 mm (2-1/2 in.) (width) between centers of adjacent runs of cable. A surface layer of gypsum board shall be mounted so that the nails or other fasteners do not pierce the heating cable.

(J) Crossing Joists. Cables shall cross joists only at the ends of the room unless the cable is required to cross joists elsewhere in order to satisfy the manufacturer's instructions that the installer avoid placing the cable too close to ceiling penetrations and luminaires (lighting fixtures).

424.42 Finished Ceilings. Finished ceilings shall not be covered with decorative panels or beams constructed of materials that have thermal insulating properties, such as wood, fiber, or plastic. Finished ceilings shall be permitted to be covered with paint, wallpaper, or other approved surface finishes.

424.43 Installation of Nonheating Leads of Cables.

(A) Free Nonheating Leads. Free nonheating leads of cables shall be installed in accordance with approved wiring methods from the junction box to a location within the ceiling. Such installations shall be permitted to be single conductors in approved raceways, single or multiconductor Type UF, Type NMC, Type MI, or other approved conductors.

(B) Leads in Junction Box. Not less than 150 mm (6 in.) of free nonheating lead shall be within the junction box. The marking of the leads shall be visible in the junction box.

(C) Excess Leads. Excess leads of heating cables shall not be cut but shall be secured to the underside of the ceiling and embedded in plaster or other approved material, leaving only a length sufficient to reach the junction box with not less than 150 mm (6 in.) of free lead within the box.

424.44 Installation of Cables in Concrete or Poured Masonry Floors.

(A) Watts per Linear Foot. Constant wattage heating cables shall not exceed 54 watts/linear meter (16-1/2 watts/linear foot) of cable.

(B) Spacing Between Adjacent Runs. The spacing between adjacent runs of cable shall not be less than 25 mm (1 in.) on centers.

(C) Secured in Place. Cables shall be secured in place by nonmetallic frames or spreaders or other approved means while the concrete or other finish is applied.

Cables shall not be installed where they bridge expansion joints unless protected from expansion and contraction.

(D) Spacings Between Heating Cable and Metal Embedded in the Floor. Spacings shall be maintained between the heating cable and metal embedded in the floor, unless the cable is a grounded metal-clad cable.

(E) Leads Protected. Leads shall be protected where they leave the floor by rigid metal conduit, intermediate metal conduit, rigid nonmetallic conduit, electrical metallic tubing, or by other approved means.

(F) Bushings or Approved Fittings. Bushings or approved fittings shall be used where the leads emerge within the floor slab.

(G) Ground-Fault Circuit-Interrupter Protection for Heated Floors of Bathrooms, and in Hydromassage Bathtub, Spa, and Hot Tub Locations. Ground-fault circuit-interrupter protection for personnel shall be provided for electrically heated floors in bathrooms, and in hydromassage bathtub, spa, and hot tub locations.

424.45 Inspection and Tests. Cable installations shall be made with due care to prevent damage to the cable assembly and shall be inspected and approved before cables are covered or concealed.

CHAPTER 22

AIR-CONDITIONING
AND REFRIGERATING EQUIPMENT

INTRODUCTION

This chapter contains provisions from Article 440—Air-Conditioning and Refrigerating Equipment. Article 440 covers electric motor–driven air-conditioning and refrigerating equipment, including their branch circuits and controllers. Provided in Article 440 are special considerations necessary for circuits supplying hermetic refrigerant motor-compressors and for any air-conditioning or refrigerating equipment that is supplied from a branch circuit that supplies a hermetic refrigerant motor-compressor.

Contained in this pocket guide are six of the seven parts that are included in Article 440. The six parts are: I General; II Disconnecting Means; III Branch-Circuit Short-Circuit and Ground-Fault Protection; IV Branch-Circuit Conductors; VI Motor-Compressor and Branch-Circuit Overload Protection; and VII Provisions for Room Air Conditioners.

ARTICLE 440
Air-Conditioning and Refrigerating Equipment

I. GENERAL

440.1 Scope. The provisions of this article apply to electric motor-driven air-conditioning and refrigerating equipment and to the branch circuits and controllers for such equipment. It provides for the special considerations necessary for circuits supplying hermetic refrigerant motor-compressors and for any air-conditioning or refrigerating equipment that is supplied from a branch circuit that supplies a hermetic refrigerant motor-compressor.

440.2 Definitions.

Branch-Circuit Selection Current. The value in amperes to be used instead of the rated-load current in determining ratings of motor branch-circuit conductors, disconnecting means, controllers, and branch-circuit short-circuit and ground-fault protective devices wherever the running overload protective device permits a sustained current greater than the specified percentage of the rated-load current. The value of branch-circuit selection current will always be equal to or greater than the marked rated-load current.

Hermetic Refrigerant Motor-Compressor. A combination consisting of a compressor and motor, both of which are enclosed in the same housing, with no external shaft or shaft seals, the motor operating in the refrigerant.

Rated-Load Current. The rated-load current for a hermetic refrigerant motor-compressor is the current resulting when the motor-compressor is operated at the rated load, rated voltage, and rated frequency of the equipment it serves.

440.3 Other Articles.

(A) Article 430. These provisions are in addition to, or amendatory of, the provisions of Article 430 and other articles in this *Code,* which apply except as modified in this article.

(B) Articles 422, 424, or 430. The rules of Articles 422, 424, or 430, as applicable, shall apply to air-conditioning and refrigerating equipment that does not incorporate a hermetic refrigerant motor-compressor. This equipment includes devices that employ refrigeration compressors driven by conventional motors, furnaces with air-conditioning evaporator coils installed, fan-coil units, remote forced air-cooled condensers, remote commercial refrigerators, and so forth.

(C) Article 422. Equipment such as room air conditioners, household refrigerators and freezers, drinking water coolers, and beverage dispensers shall be considered appliances, and the provisions of Article 422 shall also apply.

440.6 Ampacity and Rating. The size of conductors for equipment covered by this article shall be selected from Tables 310.16 through 310.19 or calculated in accordance with 310.15 as applicable. The required ampacity of conductors and rating of equipment shall be determined according to 440.6(A) and (B).

(A) Hermetic Refrigerant Motor-Compressor. For a hermetic refrigerant motor-compressor, the rated-load current marked on the nameplate of the equipment in which the motor-compressor is employed shall be used in determining the rating or ampacity of the disconnecting means, the branch-circuit conductors, the controller, the branch-circuit short-circuit and ground-fault protection, and the separate motor overload protection. Where no rated-load current is shown on the equipment nameplate, the rated-load current own on the compressor nameplate shall be used.

Exception No. 1: Where so marked, the branch-circuit selection current shall be used instead of the rated-load current to determine the rating or ampacity of the disconnecting means, the branch-circuit conductors, the controller, and the branch-circuit short-circuit and ground-fault protection.

Exception No. 2: For cord-and-plug-connected equipment, the nameplate marking shall be used in accordance with 440.22(B), Exception No. 2.

(B) Multimotor Equipment. For multimotor equipment employing a shaded-pole or permanent split-capacitor-type fan or blower motor, the full-load current for such motor marked on the nameplate of the equipment in which the fan or blower motor is employed shall be used instead of the horsepower rating to determine the ampacity or rating of the disconnecting means, the branch-circuit conductors, the controller, the branch-circuit short-circuit and ground-fault protection, and the separate overload protection. This marking on the equipment nameplate shall not be less than the current marked on the fan or blower motor nameplate.

440.8 Single Machine. An air-conditioning or refrigerating system shall be considered to be a single machine under the provisions of 430.87, Exception, and 430.112, Exception. The motors shall be permitted to be located remotely from each other.

II. DISCONNECTING MEANS

440.11 General. The provisions of Part II are intended to require disconnecting means capable of disconnecting air-conditioning and refrigerating equipment, including motor-compressors and controllers from the circuit conductors.

440.12 Rating and Interrupting Capacity.

(A) Hermetic Refrigerant Motor-Compressor. A disconnecting means serving a hermetic refrigerant motor-compressor shall be selected on the basis of the nameplate rated-load current or branch-circuit selection current, whichever is greater, and locked-rotor current, respectively, of the motor-compressor as follows:

(1) Ampere Rating. The ampere rating shall be at least 115 percent of the nameplate rated-load current or branch-circuit selection current, whichever is greater.

Exception: A listed nonfused motor circuit switch having a horsepower rating not less than the equivalent horsepower determined in accordance with 440.12(A)(2) shall be permitted have an ampere rating less than 115 percent of the specified current.

440.13 Cord-Connected Equipment. For cord-connected equipment such as room air conditioners, household refrigerators and freezers, drinking water coolers, and beverage dispensers, a separable connector or an attachment plug and receptacle shall be permitted to serve as the disconnecting means.

FPN: For room air conditioners, see 440.63.

440.14 Location. Disconnecting means shall be located within sight from and readily accessible from the air-conditioning or refrigerating equipment. The disconnecting means shall be permitted to be installed on or within the air-conditioning or refrigerating equipment.

The disconnecting means shall not be located on panels that are designed to allow access to the air-conditioning or refrigeration equipment.

III. BRANCH-CIRCUIT SHORT-CIRCUIT AND GROUND-FAULT PROTECTION

440.21 General. The provisions of Part III specify devices intended to protect the branch-circuit conductors, control apparatus, and motors in circuits supplying hermetic refrigerant motor-compressors against overcurrent due to short circuits and grounds. They are in addition to or amendatory of the provisions of Article 240.

440.22 Application and Selection.

(A) Rating or Setting for Individual Motor Compressor. The motor-compressor branch-circuit short-circuit and ground-fault protective device shall be capable of carrying the starting current of the motor. A protective device having a rating or setting not exceeding 175 percent of the motor-compressor rated-load current or branch-circuit selection current, whichever is greater, shall be permitted, provided that, where the protection specified is not sufficient for the starting current of the motor, the rating or setting shall be permitted to be increased but shall not exceed 225 percent of the motor rated-load current or branch-circuit selection current, whichever is greater.

(C) Protective Device Rating Not to Exceed the Manufacturer's Values. Where maximum protective device ratings shown on a manufacturer's overload relay table for use with a motor controller are less than the rating or setting selected in accordance with 440.22(A) and (B), the protective device rating shall not exceed the manufacturer's values marked on the equipment.

IV. BRANCH-CIRCUIT CONDUCTORS

440.31 General. The provisions of Part IV and Article 310 specify ampacities of conductors required to carry the motor current without overheating under the conditions specified, except as modified in 440.6(A), Exception No. 1.

   The provisions of these articles shall not apply to integral conductors of motors, motor controllers and the like, or to conductors that form an integral part of approved equipment.

440.32 Single Motor-Compressor. Branch-circuit conductors supplying a single motor-compressor shall have an ampacity not less than 125 percent of either the motor-compressor rated-load current or the branch-circuit selection current, whichever is greater.

440.33 Motor-Compressor(s) With or Without Additional Motor Loads. Conductors supplying one or more motor-compressor(s) with or without an additional load(s) shall have an ampacity not less than the sum of the rated-load or branch-circuit selection current ratings, whichever is larger, of all the motor-compressors plus the full-load currents of the other motors, plus 25 percent of the highest motor or motor-compressor rating in the group.

Exception No. 1: Where the circuitry is interlocked so as to prevent the starting and running of a second motor-compressor or group of motor-compressors, the conductor size shall be determined from the largest motor-compressor or group of motor-compressors that is to be operated at a given time.

Exception No. 2: The branch circuit conductors for room air conditioners shall be in accordance with Part VII of Article 440.

440.34 Combination Load. Conductors supplying a motor-compressor load in addition to a lighting or appliance load as computed from Article 220 and other applicable articles shall have an ampacity sufficient for the lighting or appliance load plus the required ampacity for the motor-compressor load determined in accordance with 440.33, or, for a single motor-compressor, in accordance with 440.32.

VI. MOTOR-COMPRESSOR AND BRANCH-CIRCUIT OVERLOAD PROTECTION

440.54 Motor-Compressors and Equipment on 15- or 20-Ampere Branch Circuits—Not Cord-and-Attachment-Plug-Connected. Overload protection for motor-compressors and equipment used on 15- or 20-ampere 120-volt, or 15-ampere 208- or 240-volt single-phase branch circuits as permitted in Article 210 shall be permitted as indicated in 440.54(A) and (B).

(A) Overload Protection. The motor-compressor shall be provided with overload protection selected as specified in 440.52(A). Both the controller and motor overload protective device shall be approved for installation with the short-circuit and ground-fault protective device for the branch circuit to which the equipment is connected.

(B) Time Delay. The short-circuit and ground-fault protective device protecting the branch circuit shall have sufficient time delay to permit the motor-compressor and other motors to start and accelerate their loads.

440.55 Cord-and-Attachment-Plug-Connected Motor-Compressors and Equipment on 15- or 20-Ampere Branch Circuits. Overload protection for motor-compressors and equipment that are cord-and-attachment-plug-connected and used on 15- or 20-ampere 120-volt, or 15-ampere 208- or 240-volt single-phase branch circuits as permitted in Article 210 shall be permitted as indicated in 440.55(A), (B), and (C).

(A) Overload Protection. The motor-compressor shall be provided with overload protection as specified in 440.52(A). Both the controller and the motor overload protective device shall be approved for installation with the short-circuit and ground-fault protective device for the branch circuit to which the equipment is connected.

(B) Attachment Plug and Receptacle Rating. The rating of the attachment plug and receptacle shall not exceed 20 amperes at 125 volts or 15 amperes at 250 volts.

(C) Time Delay. The short-circuit and ground-fault protective device protecting the branch circuit shall have sufficient time delay to permit the motor-compressor and other motors to start and accelerate their loads.

VII. PROVISIONS FOR ROOM AIR CONDITIONERS

440.60 General. The provisions of Part VII shall apply to electrically energized room air conditioners that control temperature and humidity. For the purpose of Part VII, a room air conditioner (with or without provisions for heating) shall be considered as an ac appliance of the air-cooled window, console, or in-wall type that is installed in the conditioned room and that incorporates a hermetic refrigerant motor-compressor(s). The provisions of Part VII cover equipment rated not over 250 volts, single phase, and such equipment shall be permitted to be cord-and-attachment-plug-connected.

A room air conditioner that is rated three phase or rated over 250 volts shall be directly connected to a wiring method recognized in Chapter 3, and provisions of Part VII shall not apply.

440.61 Grounding. Room air conditioners shall be grounded in accordance with 250.110, 250.112, and 250.114.

440.62 Branch-Circuit Requirements.

(A) Room Air Conditioner as a Single Motor Unit. A room air conditioner shall be considered as a single motor unit in determining its branch-circuit requirements where all the following conditions are met:

(1) It is cord-and-attachment-plug-connected.
(2) Its rating is not more than 40 amperes and 250 volts, single phase.

(3) Total rated-load current is shown on the room air-conditioner nameplate rather than individual motor currents.

(4) The rating of the branch-circuit short-circuit and ground-fault protective device does not exceed the ampacity of the branch-circuit conductors or the rating of the receptacle, whichever is less.

(B) Where No Other Loads Are Supplied. The total marked rating of a cord-and-attachment-plug-connected room air conditioner shall not exceed 80 percent of the rating of a branch circuit where no other loads are supplied.

(C) Where Lighting Units or Other Appliances Are Also Supplied. The total marked rating of a cord-and-attachment-plug-connected room air conditioner shall not exceed 50 percent of the rating of a branch circuit where lighting outlets, other appliances, or general-use receptacles are also supplied. Where the circuitry is interlocked to prevent simultaneous operation of the room air conditioner and energization of other outlets on the same branch circuit, a cord-and-attachment-plug-connected room air conditioner shall not exceed 80 percent of the branch-circuit rating.

440.63 Disconnecting Means. An attachment plug and receptacle shall be permitted to serve as the disconnecting means for a single-phase room air conditioner rated 250 volts or less if (1) the manual controls on the room air conditioner are readily accessible and located within 1.8 m (6 ft) of the floor or (2) an approved manually operable switch is installed in a readily accessible location within sight from the room air conditioner.

440.64 Supply Cords. Where a flexible cord is used to supply a room air conditioner, the length of such cord shall not exceed 3.0 m (10 ft) for a nominal, 120-volt rating or 1.8 m (6 ft) for a nominal, 208- or 240-volt rating.

CHAPTER 23

COMMUNICATIONS CIRCUITS

INTRODUCTION

This chapter contains provisions from Article 800—Communications Circuits. Article 800 covers telephone, telegraph (except radio), outside wiring for fire/burglar alarms, and similar central station systems and telephone systems not connected to a central station system but using similar types of equipment, methods of installation, and maintenance. This chapter contains two of the five parts in Article 800. Part II covers general installation requirements, and Part V covers communications wires and cables within buildings. Also covered in Part V are the listing requirements for the different cable types.

ARTICLE 800
Communications Circuits

I. GENERAL

800.1 Scope. This article covers telephone, telegraph (except radio), outside wiring for fire alarm and burglar alarm, and similar central station systems; and telephone systems not connected to a central station system but using similar types of equipment, methods of installation, and maintenance.

800.4 Equipment. Equipment intended to be electrically connected to a telecommunications network shall be listed for the purpose. Installation of equipment shall also comply with 110.3(B).

800.5 Access to Electrical Equipment Behind Panels Designed to Allow Access. Access to electrical equipment shall not be denied by an accumulation of wires and cables that prevents removal of panels, including suspended ceiling panels.

800.6 Mechanical Execution of Work. Communications circuits and equipment shall be installed in a neat and workmanlike manner. Cables installed exposed on the outer surface of ceiling and sidewalls shall be supported by the structural components of the building structure in such a manner that the cable is not be damaged by normal building use. Such cables shall be attached to structural components by straps, staples, hangers, or similar fittings designed and installed so as not to damage the cable. The installation shall also conform with 300.4(D).

IV. GROUNDING METHODS

800.40 Cable and Primary Protector Grounding. The metallic member(s) of the cable sheath, where required to be grounded by 800.33, and primary protectors shall be grounded as specified in 800.40(A) through (D).

(A) Grounding Conductor.

(1) Insulation. The grounding conductor shall be insulated and shall be listed as suitable for the purpose.

(2) Material. The grounding conductor shall be copper or other corrosion-resistant conductive material, stranded or solid.

(3) Size. The grounding conductor shall not be smaller than 14 AWG.

(4) Length. The primary protector grounding conductor shall be as short as practicable. In one- and two-family dwellings, the primary protector grounding conductor shall be as short as practicable, not to exceed 6.0 m (20 ft) in length.

Exception: In one- and two-family dwellings where it is not practicable to achieve an overall maximum primary protector grounding conductor length of 6.0 m (20 ft), a separate communications ground rod meeting the minimum dimensional criteria of 800.40(B)(3) shall be driven, the primary protector shall be grounded to the communications ground rod in accordance with 800.40(C), and the communications ground rod bonded to the power grounding electrode system in accordance with 800.40(D).

(5) Run in Straight Line. The grounding conductor shall be run to the grounding electrode in as straight a line as practicable.

(6) Physical Damage. Where necessary, the grounding conductor shall be guarded from physical damage. Where the grounding conductor is run in a metal raceway, both ends of the raceway shall be bonded to the grounding conductor or the same terminal or electrode to which the grounding conductor is connected.

(B) Electrode. The grounding conductor shall be connected in accordance with 800.40(B)(1) and (B)(2).

(1) In Buildings or Structures with Grounding Means. To the nearest accessible location on the following:

(1) The building or structure grounding electrode system as covered in 250.50
(2) The grounded interior metal water piping system, within 1.5 m (5 ft) from its point of entrance to the building, as covered in 250.52
(3) The power service accessible means external to enclosures as covered in 250.94

(4) The metallic power service raceway
(5) The service equipment enclosure
(6) The grounding electrode conductor or the grounding electrode conductor metal enclosure
(7) The grounding conductor or the grounding electrode of a building or structure disconnecting means that is grounded to an electrode as covered in 250.32.

For purposes of this section, the mobile home service equipment or the mobile home disconnecting means, as described in 800.30(B), shall be considered accessible.

(2) In Buildings or Structures Without Grounding Means. If the building or structure served has no grounding means, as described in 800.40(B)(1):

(1) To any one of the individual electrodes described in 250.52(A)(1), (2), (3), (4); or
(2) If the building or structure served has no grounding means, as described in 800.40(B)(1) or (B)(2)(1), to an effectively grounded metal structure or to a ground rod or pipe not less than 1.5 m (5 ft) in length and 12.7 mm (1/2 in.) in diameter, driven, where practicable, into permanently damp earth and separated from lightning conductors as covered in 800.13 and at least 1.8 m (6 ft) from electrodes of other systems. Steam or hot water pipes or air terminal conductors (lightning-rod conductors) shall not be employed as electrodes for protectors.

(C) Electrode Connection. Connections to grounding electrodes shall comply with 250.70. Connectors, clamps, fittings, or lugs used to attach grounding conductors and bonding jumpers to grounding electrodes or to each other that are to be concrete-encased or buried in the earth shall be suitable for its application.

(D) Bonding of Electrodes. A bonding jumper not smaller than 6 AWG copper or equivalent shall be connected between the communications grounding electrode and power grounding electrode system at the building or structure served where separate electrodes are used. Bonding together of all separate electrodes shall be permitted.

Exception: At mobile homes as covered in 800.41.

> FPN No. 1: See 250.60 for use of air terminals (lightning rods).

> FPN No. 2: Bonding together of all separate electrodes limits potential differences between them and between their associated wiring systems.

V. COMMUNICATIONS WIRES AND CABLES WITHIN BUILDINGS

800.48 Raceways for Communications Wires and Cables. Where communications wire and cables are installed in a raceway, the raceway shall be either of a type permitted in Chapter 3 and installed in accordance with Chapter 3 or a listed nonmetallic raceway complying with 800.51(J), (K), or (L), as applicable, and installed in accordance with 362.24 through 362.56, where the requirements applicable to electrical nonmetallic tubing apply.

Exception: Conduit fill restrictions shall not apply.

800.49 Fire Resistance of Communications Wires and Cables. Communications wires and cables installed as wiring within a building shall be listed as being resistant to the spread of fire in accordance with 800.50 and 800.51.

800.50 Listing, Marking, and Installation of Communications Wires and Cables. Communications wires and cables installed as wiring within buildings shall be listed as being suitable for the purpose and installed in accordance with 800.52. Communications cables and undercarpet communications wires shall be marked in accordance with Table 800.50. The cable voltage rating shall not be marked on the cable or on the undercarpet communications wire.

800.51 Listing Requirements for Communications Wires and Cables and Communications Raceways. Communications wires and cables shall have a voltage rating of not less than 300 volts and shall be listed in accordance with 800.51(A) through (J), and communications raceways shall be listed in accordance with 800.51(K) through (L). Conductors in communications cables, other than in a coaxial cable, shall be copper.

Table 800.50 Cable Markings

Cable Marking	Type	Reference
MPP	Multipurpose plenum cable	800.51(G) and 800.53(A)
CMP	Communications plenum cable	800.51(A) and 800.53(A)
MPR	Multipurpose riser cable	800.51(G) and 800.53(B)
CMR	Communications riser cable	800.51(B) and 800.53(B)
MPG	Multipurpose general-purpose cable	800.51(G) and 800.53(D) and (E)(1)
CMG	Communications general-purpose cable	800.51(C) and 800.53(D) and (E)(1)
MP	Multipurpose general-purpose cable	800.51(G) and 800.53(D) and (E)(1)
CM	Communications general-purpose cable	800.51(D) and 800.53(D) and (E)(1)
CMX	Communications cable, limited use	800.51(E) and 800.53(C), (D), and (E)
CMUC	Undercarpet communications wire and cable	800.51(F) and 800.53(F)(6)

(A) Type CMP. Type CMP communications plenum cable shall be listed as being suitable for use in ducts, plenums, and other spaces used for environmental air and shall also be listed as having adequate fire-resistant and low smoke-producing characteristics.

(B) Type CMR. Type CMR communications riser cable shall be listed as being suitable for use in a vertical run in a shaft or from floor to floor and shall also be listed as having fire-resistant characteristics capable of preventing the carrying of fire from floor to floor.

(C) Type CMG. Type CMG general-purpose communications cable shall be listed as being suitable for general-purpose communications use, with the exception of risers and plenums, and shall also be listed as being resistant to the spread of fire.

(D) Type CM. Type CM communications cable shall be listed as being suitable for general-purpose communications use, with the exception of risers and plenums, and shall also be listed as being resistant to the spread of fire.

(E) Type CMX. Type CMX limited-use communications cable shall be listed as being suitable for use in dwellings and for use in raceway and shall also be listed as being resistant to flame spread.

(F) Type CMUC UnderCarpet Wire and Cable. Type CMUC undercarpet communications wire and cable shall be listed as being suitable for undercarpet use and shall also be listed as being resistant to flame spread.

(G) Multipurpose (MP) Cables. Until July 1, 2003, cables that meet the requirements for Types CMP, CMR, CMG, and CM and also satisfy the requirements of 760.71(B) for multiconductor cables and 760.71(H) for coaxial cables shall be permitted to be listed and marked as multipurpose cable Types MPP, MPR, MPG, and MP, respectively.

(H) Communications Wires. Communications wires, such as distributing frame wire and jumper wire, shall be listed as being resistant to the spread of fire.

(I) Hybrid Power and Communications Cable. Listed hybrid power and communications cable shall be permitted where the power cable is a listed Type NM or NM-B conforming to the provisions of Article 334, and the communications cable is a listed Type CM, the jackets on the listed NM or NM-B and listed CM cables are rated for 600 volts minimum, and the hybrid cable is listed as being resistant to the spread of fire.

(J) Plenum Communications Raceways. Plenum communications raceways listed as plenum optical fiber raceways shall be permitted for use in ducts, plenums, and other spaces used for environmental air and shall also be listed as having adequate fire-resistant and low smoke-producing characteristics.

(K) Riser Communications Raceway. Riser communications raceways shall be listed as having adequate fire-resistant characteristics capable of preventing the carrying of fire from floor to floor.

(L) General-Purpose Communications Raceway. General-purpose communications raceways shall be listed as being resistant to the spread of fire.

800.52 Installation of Communications Wires, Cables, and Equipment. Communications wires and cables from the protector to the equipment or, where no protector is required, communications wires and cables attached to the outside or inside of the building shall comply with 800.52(A) through (E).

(A) Separation from Other Conductors.

(1) In Raceways, Boxes, and Cables.

(a) Other Power-Limited Circuits. Communications cables shall be permitted in the same raceway or enclosure with cables of any of the following:

(1) Class 2 and Class 3 remote-control, signaling, and power-limited circuits in compliance with Article 725
(2) Power-limited fire alarm systems in compliance with Article 760
(3) Nonconductive and conductive optical fiber cables in compliance with Article 770
(4) Community antenna television and radio distribution systems in compliance with Article 820
(5) Low-power network-powered broadband communications circuits in compliance with Article 830

(b) Class 2 and Class 3 Circuits. Class 1 circuits shall not be run in the same cable with communications circuits. Class 2 and Class 3 circuit conductors shall be permitted in the same cable with communications circuits, in which case the Class 2 and Class 3 circuits shall be classified as communications circuits and shall meet the requirements of this article. The cables shall be listed as communications cables or multipurpose cables.

(c) Electric Light, Power, Class 1, Non–Power-Limited Fire Alarm, and Medium Power Network-Powered Broadband Communications Circuits in Raceways, Compartments, and Boxes. Communications conductors shall not be placed in any raceway, compartment, outlet box, junction box, or similar fitting with conductors of electric light, power, Class 1, non–power-limited fire alarm or medium power network-powered broadband communications circuits.

Exception No. 1: Where all of the conductors of electric light, power, Class 1, non–power-limited fire alarm, and medium power network-powered broadband communications circuits are separated from all of the conductors of communications circuits by a barrier.

Exception No. 2: Power conductors in outlet boxes, junction boxes, or similar fittings or compartments where such conductors are introduced solely for power supply to communications equipment. The power circuit conductors shall be routed within the enclosure to maintain a minimum of 6 mm (0.25 in.) separation from the communications circuit conductors.

(B) Spread of Fire or Products of Combustion. Installations in hollow spaces, vertical shafts, and ventilation or air-handling ducts shall be made so that the possible spread of fire or products of combustion is not substantially increased. Openings around penetrations through fire resistance-rated walls, partitions, floors, or ceilings shall be fire-stopped using approved methods to maintain the fire resistance rating.

The accessible portion of abandoned communications cables shall not be permitted to remain.

(E) Support of Conductors. Raceways shall be used for their intended purpose. Communications cables or wires shall not be strapped, taped, or attached by any means to the exterior of any conduit or raceway as a means of support.

CHAPTER 24

SEPARATE BUILDINGS OR STRUCTURES

INTRODUCTION

This chapter contains provisions from two sections of Article 225—Outside Branch Circuits and Feeders and from one section of Article 250—Grounding. This chapter covers the section pertaining to two or more buildings or structures supplied from a common service. This section's requirements cover: grounding electrode(s), grounded systems, disconnecting means located in a separate building or structure on the same premises, and the grounding electrode conductor.

ARTICLE 225
Outside Branch Circuits

225.31 Disconnecting Means. Means shall be provided for disconnecting all ungrounded conductors that supply or pass through the building or structure.

225.32 Location. The disconnecting means shall be installed either inside or outside of the building or structure served or where the conductors pass through the building or structure. The disconnecting means shall be at a readily accessible location nearest the point of entrance of the conductors. For the purposes of this section, the requirements in 230.6 shall be permitted to be utilized.

Exception No. 1: For installations under single management, where documented safe switching procedures are established and maintained for disconnection, and where the installation is monitored by qualified individuals, the disconnecting means shall be permitted to be located elsewhere on the premises.

Exception No. 2: For buildings or other structures qualifying under the provisions of Article 685, the disconnecting means shall be permitted to be located elsewhere on the premises.

Exception No. 3: For towers or poles used as lighting standards, the disconnecting means shall be permitted to be located elsewhere on the premises.

Exception No. 4: For poles or similar structures used only for support of signs installed in accordance with Article 600, the disconnecting means shall be permitted to be located elsewhere on the premises.

ARTICLE 250
Grounding

250.32 Two or More Buildings or Structures Supplied from a Common Service.

(A) Grounding Electrode. Where two or more buildings or structures are supplied from a common ac service by a feeder(s) or branch circuit(s), the grounding electrode(s) required in Part III of this article at each building or structure shall be connected in the manner specified in 250.32(B) or (C). Where there are no existing grounding electrodes, the grounding electrode(s) required in Part III of this article shall be installed.

Exception: A grounding electrode at separate buildings or structures shall not be required where only one branch circuit supplies the building or structure and the branch circuit includes an equipment grounding conductor for grounding the conductive non–current-carrying parts of all equipment.

(B) Grounded Systems. For a grounded system at the separate building or structure, the connection to the grounding electrode and grounding or bonding of equipment, structures, or frames required to be grounded or bonded shall comply with either 250.32(B)(1) or (2).

(1) Equipment Grounding Conductor. An equipment grounding conductor as described in 250.118 shall be run with the supply conductors and connected to the building or structure disconnecting means and to the grounding electrode(s). The equipment grounding conductor shall be used for grounding or bonding of equipment, structures, or frames required to be grounded or bonded. The equipment grounding conductor shall be sized in accordance with 250.122. Any installed grounded conductor shall not be connected to the equipment grounding conductor or to the grounding electrode(s).

(2) Grounded Conductor. Where (1) an equipment grounding conductor is not run with the supply to the building or structure, (2) there are no continuous metallic paths bonded to the grounding system in both buildings or structures involved, and (3) ground-fault protection of equipment has not been installed on the common ac service, the grounded circuit conductor run with the supply to the building or structure shall be connected to the building or structure disconnecting means and to the grounding electrode(s) and shall be used for grounding or bonding of equipment, structures, or frames required to be grounded or bonded. The size of the grounded conductor shall not be smaller than the larger of

(1) That required by 220.22
(2) That required by 250.122

(D) Disconnecting Means Located in Separate Building or Structure on the Same Premises. Where one or more disconnecting means supply one or more additional buildings or structures under single management, and where these disconnecting means are located remote from those buildings or structures in accordance with the provisions of 225.32, Exception Nos. 1 and 2, all of the following conditions shall be met:

(1) The connection of the grounded circuit conductor to the grounding electrode at a separate building or structure shall not be made.
(2) An equipment grounding conductor for grounding any non–current-carrying equipment, interior metal piping systems, and building or structural metal frames is run with the circuit conductors to a separate building or structure and bonded to existing grounding electrode(s) required in Part III of this article, or, where there are no existing electrodes, the grounding electrode(s) required in Part III of this article shall be installed where a separate building or structure is supplied by more than one branch circuit.
(3) Bonding the equipment grounding conductor to the grounding electrode at a separate building or structure shall be made in a junction box, panelboard, or similar enclosure located immediately inside or outside the separate building or structure.

(E) Grounding Electrode Conductor. The size of the grounding electrode conductor to the grounding electrode(s) shall not be smaller than given in 250.66, based on the largest ungrounded supply conductor. The installation shall comply with Part III of this article.

CHAPTER 25

SWIMMING POOLS, FOUNTAINS, AND SIMILAR INSTALLATIONS

INTRODUCTION

This chapter contains provisions from Article 680—Swimming Pools, Fountains, and Similar Installations. The provisions of this article apply to the construction and installation of electric wiring for and equipment in (or adjacent to) all swimming, wading, and decorative pools, fountains, hot tubs, spas, and hydromassage bathtubs, whether permanently installed or storable, as well as to metallic auxiliary equipment (such as pumps, filters, etc.). The requirements of Article 680 pertain to grounding, bonding, conductor clearances, receptacles, lighting, equipment, and ground-fault circuit-interrupter protection for personnel.

Contained in this pocket guide are six of the seven parts that are included in Article 680. The six parts are: I General; II Permanently Installed Pools; III Storable Pools; IV Spas and Hot Tubs; V Fountains; and VII Hydromassage Bathtubs.

ARTICLE 680
Swimming Pools, Fountains, and Similar Installations

I. GENERAL

680.1 Scope. The provisions of this article apply to the construction and installation of electrical wiring for and equipment in or adjacent to all swimming, wading, therapeutic, and decorative pools, fountains, hot tubs, spas, and hydromassage bathtubs, whether permanently installed or storable, and to metallic auxiliary equipment, such as pumps, filters, and similar equipment. The term *body of water* used throughout Part I applies to all bodies of water covered in this scope unless otherwise amended.

680.6 Grounding. Electrical equipment shall be grounded in accordance with Parts V, VI, and VII of Article 250 and connected by wiring methods of Chapter 3, except as modified by this article. The following equipment shall be grounded:

(1) Through-wall lighting assemblies and underwater luminaires (lighting fixtures), other than those low-voltage systems listed for the application without a grounding conductor
(2) All electrical equipment located within 1.5 m (5 ft) of the inside wall of the specified body of water
(3) All electrical equipment associated with the recirculating system of the specified body of water
(4) Junction boxes
(5) Transformer enclosures
(6) Ground-fault circuit interrupters
(7) Panelboards that are not part of the service equipment and that supply any electrical equipment associated with the specified body of water

680.8 Overhead Conductor Clearances.

(A) Power. With respect to service drop conductors and open overhead wiring, swimming pool and similar installations shall comply with the minimum clearances given in Table 680.8 and illustrated in Figure 680.8.

Table 660.8 Overhead Conductor Clearances

Clearance Parameters	Insulated Cables, 0–750 Volts to Ground, Supported on and Cabled Together with an Effectively Grounded Bare Messenger or Effectively Grounded Neutral Conductor		All Other Conductors Voltage to Ground			
			0 through 15 kV		Over 15 through 50 kV	
	m	ft	m	ft	m	ft
A. Clearance in any direction to the water level, edge of water surface, base of diving platform, or permanently anchored raft	6.9	22.5	7.5	25	8.0	27
B. Clearance in any direction to the observation stand, tower, or diving platform	4.4	14.5	5.2	17	5.5	18
C. Horizontal limit of clearance measured from inside wall of the pool	This limit shall extend to the outer edge of the structures listed in A and B of this table but not less than 3 m (10 ft).					

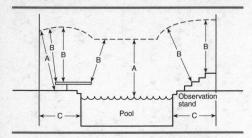

Figure 680.8 Clearances from pool structures.

FPN: Open overhead wiring as used in this article typically refers to conductor(s) not in an enclosed raceway.

680.10 Underground Wiring Location. Underground wiring shall not be permitted under the pool or within the area extending 1.5 m (5 ft) horizontally from the inside wall of the pool unless this wiring is necessary to supply pool equipment permitted by this article. Where space limitations prevent wiring from being routed a distance 1.5 m (5 ft) or more from the pool, such wiring shall be permitted where installed in rigid metal conduit, intermediate metal conduit, or a nonmetallic raceway system. All metal conduit shall be corrosion resistant and suitable for the location. The minimum burial depth shall be as given in Table 680.10.

680.12 Maintenance Disconnecting Means. One or more means to disconnect all ungrounded conductors shall be provided for all utilization equipment other than lighting. Each means shall be accessible and within sight from its equipment.

II. PERMANENTLY INSTALLED POOLS

680.20 General. Electrical installations at permanently installed pools shall comply with the provisions of Part I and Part II of this article.

Table 680.10 Minimum Burial Depths

Wiring Method	Minimum Burial	
	mm	in.
Rigid metal conduit	150	6
Intermediate metal conduit	150	6
Nonmetallic raceways listed for direct burial without concrete encasement	450	18
Other approved raceways*	450	18

*Raceways approved for burial only where concrete encased shall require a concrete envelope not less than 50 mm (2 in.) thick.

680.21 Motors.

(A) Wiring Methods.

(1) General. The branch circuits for pool-associated motors shall be installed in rigid metal conduit, intermediate metal conduit, rigid nonmetallic conduit, or Type MC cable listed for the location. Other wiring methods and materials shall be permitted in specific locations or applications as covered in this section. Any wiring method employed shall contain a copper equipment grounding conductor sized in accordance with 250.122 but not smaller than 12 AWG.

(2) On or Within Buildings. Where installed on or within buildings, electrical metallic tubing shall be permitted.

(3) Flexible Connections. Where necessary to employ flexible connections at or adjacent to the motor, liquidtight flexible metal or nonmetallic conduit with approved fittings shall be permitted.

(4) One-Family Dwellings. In the interior of one-family dwellings, or in the interior of accessory buildings associated with a one-family dwelling, any of the wiring methods recognized in Chapter 3 of this *Code* shall be permitted that comply with the provisions of this paragraph. Where run in a raceway, the equipment grounding conductor shall be insulated. Where run in a cable assembly, the equipment grounding conductor shall be permitted to be uninsulated, but it shall be enclosed within the outer sheath of the cable assembly.

(5) Cord-and-Plug Connections. Pool-associated motors shall be permitted to employ cord-and-plug connections. The flexible cord shall not exceed 900 mm (3 ft) in length. The flexible cord shall include an equipment grounding conductor sized in accordance with 250.122 and shall terminate in a grounding-type attachment plug.

(B) Double Insulated Pool Pumps. A listed cord-and-plug-connected pool pump incorporating an approved system of double insulation that provides a means for grounding only the internal and nonaccessible, non–current-carrying metal parts of the pump shall be connected to any wiring method recognized in Chapter 3 that is suitable for the location.

680.22 Area Lighting, Receptacles, and Equipment.

(A) Receptacles.

(1) Circulation and Sanitation System, Location. Receptacles that provide power for water-pump motors or for other loads directly related to the circulation and sanitation system shall be located at least 3.0 m (10 ft) from the inside walls of the pool, or not less than 1.5 m (5 ft) from the inside walls of the pool if they meet all of the following conditions:

(1) Consist of single receptacles
(2) Employ a locking configuration
(3) Are of the grounding type
(4) Have GFCI protection

(2) Other Receptacles, Location. Other receptacles shall be not less than 3.0 m (10 ft) from the inside walls of a pool.

(3) Dwelling Unit(s). Where a permanently installed pool is installed at a dwelling unit(s), no fewer than one 125-volt 15- or 20-ampere receptacle on a general-purpose branch circuit shall be located not less than 3.0 m (10 ft) from and not more than 6.0 m (20 ft) from the inside wall of the pool. This receptacle shall be located not more than 2.0 m (6 ft 6 in.) above the floor, platform, or grade level serving the pool.

(4) Restricted Space. Where a pool is within 3.0 m (10 ft) of a dwelling and the dimensions of the lot preclude meeting the required clearances, not more than one receptacle outlet shall be permitted if not less than 1.5 m (5 ft) measured horizontally from the inside wall of the pool.

(5) GFCI Protection. All 125-volt receptacles located within 6.0 m (20 ft) of the inside walls of a pool or fountain shall be protected by a ground-fault circuit interrupter. Receptacles that supply pool pump motors and that are rated 15 or 20 amperes, 120 volt through 240 volts, single phase, shall be provided with GFCI protection.

(6) Measurements. In determining the dimensions in this section addressing receptacle spacings, the distance to be measured shall be the shortest path the supply cord of an appliance connected to the receptacle would follow without piercing a floor, wall, ceiling, doorway with hinged or sliding door, window opening, or other effective permanent barrier.

(B) Luminaires (Lighting Fixtures), Lighting Outlets, and Ceiling-Suspended (Paddle) Fans.

(1) New Outdoor Installation Clearances. In outdoor pool areas, luminaires (lighting fixtures), lighting outlets, and ceiling-suspended (paddle) fans installed above the pool or the area extending 1.5 m (5 ft) horizontally from the inside walls of the pool shall be installed at a height not less than 3.7 m (12 ft) above the maximum water level of the pool.

(2) Indoor Clearances. For installations in indoor pool areas, the clearances shall be the same as for outdoor areas unless modified as provided in this paragraph. If the branch circuit supplying the equipment is protected by a ground-fault circuit interrupter, the following equipment shall be permitted at a height not less than 2.3 m (7 ft 6 in.) above the maximum pool water level:

(1) Totally enclosed luminaires (fixtures)
(2) Ceiling-suspended (paddle) fans identified for use beneath ceiling structures such as provided on porches or patios

(3) Existing Installations. Existing luminaires (lighting fixtures) and lighting outlets located less than 1.5 m (5 ft) measured horizontally from the inside walls of a pool shall be not less than 1.5 m (5 ft) above the surface of the maximum water level, shall be rigidly attached to the existing structure, and shall be protected by a ground-fault circuit interrupter.

(4) GFCI Protection in Adjacent Areas. Luminaires (lighting fixtures), lighting outlets, and ceiling-suspended (paddle) fans installed in the area extending between 1.5 m (5 ft) and 3.0 m (10 ft) horizontally from the inside walls of a pool shall be protected by a ground-fault circuit interrupter unless installed not less than 1.5 m (5 ft) above the maximum water level and rigidly attached to the structure adjacent to or enclosing the pool.

(5) Cord-and-Plug-Connected Luminaires (Lighting Fixtures). Cord-and-plug-connected luminaires (lighting fixtures) shall comply with the requirements of 680.7 where installed within 4.9 m (16 ft) of any point on the water surface, measured radially.

(C) Switching Devices. Switching devices shall be located at least 1.5 m (5 ft) horizontally from the inside walls of a pool unless separated from the pool by a solid fence, wall, or other permanent barrier. Alternatively, a switch that is listed as being acceptable for use within 1.5 m (5 ft) shall be permitted.

680.23 Underwater Luminaires (Lighting Fixtures). This section covers all luminaires (lighting fixtures) installed below the normal water level of the pool.

(A) General.

(1) Luminaire (Fixture) Design, Normal Operation. The design of an underwater luminaire (lighting fixture) supplied from a branch circuit either directly or by way of a transformer meeting the requirements of this section shall be such that, where the luminaire (fixture) is properly installed without a ground-fault circuit interrupter, there is no shock hazard with any likely combination of fault conditions during normal use (not relamping).

(2) Transformers. Transformers used for the supply of underwater luminaires (fixtures), together with the transformer enclosure, shall be listed for the purpose. The transformer shall be an isolated winding type with an ungrounded secondary that has a grounded metal barrier between the primary and secondary windings.

(3) GFCI Protection, Relamping. A ground-fault circuit interrupter shall be installed in the branch circuit supplying luminaires (fixtures) operating at more than 15 volts, so that there is no shock hazard during relamping. The installation of the ground-fault circuit interrupter shall be such that there is no shock hazard with any likely fault-condition combination that involves a person in a conductive path from any ungrounded part of the branch circuit or the luminaire (fixture) to ground.

(4) Voltage Limitation. No luminaires (lighting fixtures) shall be installed for operation on supply circuits over 150 volts between conductors.

(5) Location, Wall-Mounted Luminaires (Fixtures). Luminaires (lighting fixtures) mounted in walls shall be installed with the top of the luminaire (fixture) lens not less than 450 mm (18 in.) below the normal water level of the pool, unless the luminaire (lighting fixture) is listed and identified for use at lesser depths. No luminaire (fixture) shall be installed less than 100 mm (4 in.) below the normal water level of the pool.

(6) Bottom-Mounted Luminaires (Fixtures). A luminaire (lighting fixture) facing upward shall have the lens adequately guarded to prevent contact by any person.

(7) Dependence on Submersion. Luminaires (fixtures) that depend on submersion for safe operation shall be inherently protected against the hazards of overheating when not submerged.

(8) Compliance. Compliance with these requirements shall be obtained by the use of a listed underwater luminaire (lighting fixture) and by installation of a listed ground-fault circuit interrupter in the branch circuit or a listed transformer for luminaires (fixtures) operating at not more than 15 volts.

(B) Wet-Niche Luminaires (Fixtures).

(1) Forming Shells. Forming shells shall be installed for the mounting of all wet-niche underwater luminaires (fixtures) and shall be equipped with provisions for conduit entries. Metal parts of the luminaire (fixture) and forming shell in contact with the pool water shall be of brass or other approved corrosion-resistant metal. All forming shells used with nonmetallic conduit systems, other than those that are part of a listed low-voltage lighting system not requiring grounding, shall include provisions for terminating an 8 AWG copper conductor.

(2) Wiring Extending Directly to the Forming Shell. Conduit shall be installed from the forming shell to a suitable junction box or other enclosure located as provided in 680.24. Conduit shall be rigid metal, intermediate metal, liquidtight flexible nonmetallic, or rigid nonmetallic.

(a) Metal Conduit. Metal conduit shall be approved and shall be of brass or other approved corrosion-resistant metal.

(b) Nonmetallic Conduit. Where a nonmetallic conduit is used, an 8 AWG insulated solid or stranded copper equipment grounding conductor shall be installed in this conduit unless a listed low-voltage lighting system not requiring grounding is used. The equipment grounding conductor shall be terminated in the forming shell, junction box or transformer enclosure, or ground-fault circuit-interrupter enclosure. The termination of the 8 AWG equipment grounding conductor in the forming shell shall be covered with, or encapsulated in, a listed potting compound to protect the connection from the possible deteriorating effect of pool water.

(3) Equipment Grounding Provisions for Cords. Wet-niche luminaires (lighting fixtures) that are supplied by a flexible cord or cable shall have all exposed non–current-carrying metal parts grounded by an insulated copper equipment grounding conductor that is an integral part of the cord or cable. This grounding conductor shall be connected to a grounding terminal in the supply junction box, transformer enclosure, or other enclosure. The grounding conductor shall not be smaller than the supply conductors and not smaller than 16 AWG.

(4) Luminaire (Fixture) Grounding Terminations. The end of the flexible-cord jacket and the flexible-cord conductor terminations within a luminaire (fixture) shall be covered with, or encapsulated in, a suitable potting compound to prevent the entry of water into the luminaire (fixture) through the cord or its conductors. In addition, the grounding connection within a luminaire (fixture) shall be similarly treated to protect such connection from the deteriorating effect of pool water in the event of water entry into the luminaire (fixture).

(5) Luminaire (Fixture) Bonding. The luminaire (fixture) shall be bonded to and secured to the forming shell by a positive locking device that ensures a low-resistance contact and requires a tool to remove the luminaire (fixture) from the forming shell. Bonding shall not be required for luminaires (fixtures) that are listed for the application and have no non–current-carrying metal parts.

(C) Dry-Niche Luminaires (Fixtures).

(1) Construction. A dry-niche luminaire (lighting fixture) shall be provided with a provision for drainage of water and a means for accommodating one equipment grounding conductor for each conduit entry.

(2) Junction Box. A junction box shall not be required but, if used, shall not be required to be elevated or located as specified in 680.24(A)(2) if the luminaire (fixture) is specifically identified for the purpose.

(D) No-Niche Luminaires (Fixtures). A no-niche luminaire (fixture) shall meet the construction requirements of 680.23(B)(3) and be installed in accordance with the requirements of 680.23(B). Where connection to a forming shell is specified, the connection shall be to the mounting bracket.

(E) Through-Wall Lighting Assembly. A through-wall lighting assembly shall be equipped with a threaded entry or hub, or a nonmetallic hub listed for the purpose, for the purpose of accommodating the termination of the supply conduit. A through-wall lighting assembly shall meet the construction requirements of 680.23(B)(3) and be installed in accordance with the requirements of 680.23. Where connection to a forming shell is specified, the connection shall be to the conduit termination point.

(F) Branch-Circuit Wiring.

(1) Wiring Methods. Branch-circuit wiring on the supply side of enclosures and junction boxes connected to conduits run to wet-niche and no-niche luminaires (fixtures), and the field wiring compartments of dry-niche luminaires (fixtures), shall be installed using rigid metal conduit, intermediate metal conduit, liquidtight flexible nonmetallic conduit, or rigid nonmetallic conduit. Where installed on buildings, electrical metallic tubing shall be permitted, and where installed within buildings, electrical nonmetallic tubing or electrical metallic tubing shall be permitted.

Exception: Where connecting to transformers for pool lights, liquidtight flexible metal conduit or liquidtight flexible nonmetallic conduit shall be permitted. The length shall not exceed 1.8 m (6 ft) for any one length or exceed 3.0 m (10 ft) in total length used. Liquidtight flexible nonmetallic conduit, Type B (LFNC-B), shall be permitted in lengths longer than 1.8 m (6 ft).

(2) Equipment Grounding. Through-wall lighting assemblies, wet-niche, dry-niche, or no-niche luminaires (lighting fixtures) shall be connected to an insulated copper equipment grounding conductor installed with the circuit conductors. The equipment grounding conductor shall be installed without joint or splice except as permitted in (a) and (b). The equipment grounding conductor shall be sized in accordance with Table 250.122 but shall not be smaller than 12 AWG.

Exception: An equipment grounding conductor between the wiring chamber of the secondary winding of a transformer and a junction box shall be sized in accordance with the overcurrent device in this circuit.

(a) If more than one underwater luminaire (lighting fixture) is supplied by the same branch circuit, the equipment grounding conductor, installed between the junction boxes, transformer enclosures, or other enclosures in the supply circuit to wet-niche luminaires (fixtures), or between the field-wiring compartments of dry-niche luminaires (fixtures), shall be permitted to be terminated on grounding terminals.

(b) If the underwater luminaire (lighting fixture) is supplied from a transformer, ground-fault circuit interrupter, clock-operated switch, or a manual snap switch that is located between the panelboard and a junction box connected to the conduit that extends directly to the underwater luminaire (lighting fixture), the equipment grounding conductor shall be permitted to terminate on grounding terminals on the transformer, ground-fault circuit interrupter, clock-operated switch enclosure, or an outlet box used to enclose a snap switch.

(3) Conductors. Conductors on the load side of a ground-fault circuit interrupter or of a transformer, used to comply with the provisions of 680.23(A)(8), shall not occupy raceways, boxes, or enclosures containing other conductors unless one of the following conditions applies:

(1) The other conductors are protected by ground-fault circuit interrupters.
(2) The other conductors are grounding conductors.
(3) The other conductors are supply conductors to a feed-through type ground-fault circuit interrupter.
(4) Ground-fault circuit interrupters shall be permitted in a panelboard that contains circuits protected by other than ground-fault circuit interrupters.

680.24 Junction Boxes and Enclosures for Transformers or Ground-Fault Circuit Interrupters.

(A) Junction Boxes. A junction box connected to a conduit that extends directly to a forming shell or mounting bracket of a no-niche luminaire (fixture) shall meet the requirements of this section.

(1) Construction. The junction box shall be listed and labeled for the purpose and shall comply with the following conditions:

(1) Be equipped with threaded entries or hubs or a nonmetallic hub listed for the purpose

(2) Be comprised of copper, brass, suitable plastic, or other approved corrosion-resistant material

(3) Be provided with electrical continuity between every connected metal conduit and the grounding terminals by means of copper, brass, or other approved corrosion-resistant metal that is integral with the box

(2) Installation. Where the luminaire (fixture) operates over 15 volts, the junction box location shall comply with (a) and (b). Where the luminaire (fixture) operates at less than 15 volts, the junction box location shall be permitted to comply with (c).

(a) Vertical Spacing. The junction box shall be located not less than 100 mm (4 in.), measured from the inside of the bottom of the box, above the ground level, or pool deck, or not less than 200 mm (8 in.) above the maximum pool water level, whichever provides the greater elevation.

(b) Horizontal Spacing. The junction box shall be located not less than 1.2 m (4 ft) from the inside wall of the pool, unless separated from the pool by a solid fence, wall, or other permanent barrier.

(c) Flush Deck Box. If used on a lighting system operating at 15 volts or less, a flush deck box shall be permitted if both of the following apply:

(1) An approved potting compound is used to fill the box to prevent the entrance of moisture.

(2) The flush deck box is located not less than 1.2 m (4 ft) from the inside wall of the pool.

(B) Other Enclosures. An enclosure for a transformer, ground-fault circuit interrupter, or a similar device connected to a conduit that extends directly to a forming shell or mounting bracket of a no-niche luminaire (fixture) shall meet the requirements of this section.

(1) Construction. The enclosure shall be listed and labeled for the purpose and meet the following requirements:

(1) Equipped with threaded entries or hubs or a nonmetallic hub listed for the purpose

(2) Comprised of copper, brass, suitable plastic, or other approved corrosion-resistant material

(3) Provided with an approved seal, such as duct seal at the conduit connection, that prevents circulation of air between the conduit and the enclosures

(4) Provided with electrical continuity between every connected metal conduit and the grounding terminals by means of copper, brass, or other approved corrosion-resistant metal that is integral with the box

(2) Installation.

(a) Vertical Spacing. The enclosure shall be located not less than 100 mm (4 in.), measured from the inside of the bottom of the box, above the ground level, or pool deck, or not less than 200 mm (8 in.) above the maximum pool water level, whichever provides the greater elevation.

(b) Horizontal Spacing. The enclosure shall be located not less than 1.2 m (4 ft) from the inside wall of the pool, unless separated from the pool by a solid fence, wall, or other permanent barrier.

(C) Protection. Junction boxes and enclosures mounted above the grade of the finished walkway around the pool shall not be located in the walkway unless afforded additional protection, such as by location under diving boards, adjacent to fixed structures, and the like.

(D) Grounding Terminals. Junction boxes, transformer enclosures, and ground-fault circuit-interrupter enclosures connected to a conduit that extends directly to a forming shell or mounting bracket of a no-niche luminaire (fixture) shall be provided with a number of grounding terminals that shall be no fewer than one more than the number of conduit entries.

(E) Strain Relief. The termination of a flexible cord of an underwater luminaire (lighting fixture) within a junction box, transformer enclosure, ground-fault circuit interrupter, or other enclosure shall be provided with a strain relief.

(F) Grounding. The junction box, transformer enclosure, or other enclosure in the supply circuit to a wet-niche or no-niche luminaire (lighting fixture) and the field-wiring chamber of a dry-niche luminaire (lighting fixture) shall be grounded to the equipment grounding terminal of the panelboard. This terminal shall be directly connected to the panelboard enclosure.

680.25 Feeders. These provisions shall apply to any feeder on the supply side of panelboards supplying branch circuits for pool equipment covered in Part II of this article and on the load side of the service equipment or the source of a separately derived system.

(A) Wiring Methods. Feeders shall be installed in rigid metal conduit, intermediate metal conduit, liquidtight flexible nonmetallic conduit, or rigid nonmetallic conduit. Electrical metallic tubing shall be permitted where installed on or within a building, and electrical nonmetallic tubing shall be permitted where installed within a building.

Exception: An existing feeder between an existing remote panelboard and service equipment shall be permitted to run in flexible metal conduit or an approved cable assembly that includes an equipment grounding conductor within its outer sheath. The equipment grounding conductor shall comply with 250.24(A)(5).

(B) Grounding. An equipment grounding conductor shall be installed with the feeder conductors between the grounding terminal of the pool equipment panelboard and the grounding terminal of the applicable service equipment or source of a separately derived system. For other than (1) existing feeders covered in 680.25(A), Exception or (2) feeders to separate buildings that do not utilize an insulated equipment grounding conductor in accordance with 680.25(B)(2), this equipment grounding conductor shall be insulated.

(1) Size. This conductor shall be sized in accordance with 250.122 but not smaller than 12 AWG. On separately derived systems, this conductor shall be sized in accordance with Table 250.66 but not smaller than 8 AWG.

(2) Separate Buildings. A feeder to a separate building shall be permitted to supply swimming pool equipment branch circuits, or feeders supplying swimming pool equipment branch circuits, if the grounding arrangements in the separate building meet the requirements in 250.32. Where installed, a separate equipment grounding conductor shall be an insulated conductor.

680.26 Bonding.

(A) Performance. The bonding required by this section shall be installed to eliminate voltage gradients in the pool area as prescribed.

> FPN: This section does not require that the 8 AWG or larger solid copper bonding conductor be extended or attached to any remote panelboard, service equipment, or any electrode.

(B) Bonded Parts. The parts specified in 680.26(B)(1) through (B)(5) shall be bonded together.

(1) Metallic Structural Components. All metallic parts of the pool structure, including the reinforcing metal of the pool shell, coping stones, and deck, shall be bonded. The usual steel tie wires shall be considered suitable for bonding the reinforcing steel together, and welding or special clamping shall not be required. These tie wires shall be made tight. If reinforcing steel is effectively insulated by an encapsulating nonconductive compound at the time of manufacture and installation, it shall not be required to be bonded. Where reinforcing steel is encapsulated with a nonconductive compound, provisions shall be made for an alternate means to eliminate voltage gradients that would otherwise be provided by unencapsulated, bonded reinforcing steel.

(2) Underwater Lighting. All forming shells and mounting brackets of no-niche luminaires (fixtures) shall be bonded unless a listed low-voltage lighting system with nonmetallic forming shells not requiring bonding is used.

(3) Metal Fittings. All metal fittings within or attached to the pool structure shall be bonded. Isolated parts that are not over 100 mm (4 in.) in any dimension and do not penetrate into the pool structure more than 25 mm (1 in.) shall not require bonding.

(4) Electrical Equipment. Metal parts of electrical equipment associated with the pool water circulating system, including pump motors and metal parts of equipment associated with pool covers, including electric motors, shall be bonded. Metal parts of listed equipment incorporating an approved system of double insulation and providing a means for grounding internal nonaccessible, non–current-carrying metal parts shall not be bonded.

Where a double-insulated water-pump motor is installed under the provisions of this rule, a solid 8 AWG copper conductor that is of sufficient length to make a bonding connection to a replacement motor shall be extended from the bonding grid to an accessible point in the motor vicinity. Where there is no connection between the swimming pool bonding grid and the equipment grounding system for the premises, this bonding conductor shall be connected to the equipment grounding conductor of the motor circuit.

(5) Metal Wiring Methods and Equipment. Metal-sheathed cables and raceways, metal piping, and all fixed metal parts except those separated from the pool by a permanent barrier shall be bonded that are within the following distances of the pool:

(1) Within 1.5 m (5 ft) horizontally of the inside walls of the pool
(2) Within 3.7 m (12 ft) measured vertically above the maximum water level of the pool, or any observation stands, towers, or platforms, or any diving structures

(C) Common Bonding Grid. The parts specified in 680.26(B) shall be connected to a common bonding grid with a solid copper conductor, insulated, covered, or bare, not smaller than 8 AWG. Connection shall be made by exothermic welding or by pressure connectors or clamps that are labeled as being suitable for the purpose and are of stainless steel, brass, copper, or copper alloy. The common bonding grid shall be permitted to be any of the following:

(1) The structural reinforcing steel of a concrete pool where the reinforcing rods are bonded together by the usual steel tie wires or the equivalent
(2) The wall of a bolted or welded metal pool
(3) A solid copper conductor, insulated, covered, or bare, not smaller than 8 AWG

(4) Rigid metal conduit or intermediate metal conduit of brass or other identified corrosion-resistant metal conduit

(D) Connections. Where structural reinforcing steel or the walls of bolted or welded metal pool structures are used as a common bonding grid for nonelectrical parts, the connections shall be made in accordance with 250.8.

(E) Pool Water Heaters. For pool water heaters rated at more than 50 amperes that have specific instructions regarding bonding and grounding, only those parts designated to be bonded shall be bonded, and only those parts designated to be grounded shall be grounded.

680.27 Specialized Pool Equipment.

(A) Underwater Audio Equipment. All underwater audio equipment shall be identified for the purpose.

(B) Electrically Operated Pool Covers.

(1) Motors and Controllers. The electric motors, controllers, and wiring shall be located not less than 1.5 m (5 ft) from the inside wall of the pool unless separated from the pool by a wall, cover, or other permanent barrier. Electric motors installed below grade level shall be of the totally enclosed type. The device that controls the operation of the motor for an electrically operated pool cover shall be located so that the operator has full view of the pool.

(2) Protection. The electric motor and controller shall be connected to a circuit protected by a ground-fault circuit interrupter.

(C) Deck Area Heating. These provisions of this section shall apply to all pool deck areas, including a covered pool, where electrically operated comfort heating units are installed within 6.0 m (20 ft) of the inside wall of the pool.

(1) Unit Heaters. Unit heaters shall be rigidly mounted to the structure and shall be of the totally enclosed or guarded types. Unit heaters shall not be mounted over the pool or within the area extending 1.5 m (5 ft) horizontally from the inside walls of a pool.

(2) Permanently Wired Radiant Heaters. Radiant electric heaters shall be suitably guarded and securely fastened to their mounting device(s). Heaters shall not be installed over a pool or within the area extending 1.5 m (5 ft) horizontally from the inside walls of the pool and shall be mounted at least 3.7 m (12 ft) vertically above the pool deck unless otherwise approved.

(3) Radiant Heating Cables Not Permitted. Radiant heating cables embedded in or below the deck shall not be permitted.

III. STORABLE POOLS

680.30 General. Electrical installations at storable pools shall comply with the provisions of Part I and Part III of this article.

680.31 Pumps. A cord-connected pool filter pump shall incorporate an approved system of double insulation or its equivalent and shall be provided with means for grounding only the internal and nonaccessible non–current-carrying metal parts of the appliance.

The means for grounding shall be an equipment grounding conductor run with the power-supply conductors in the flexible cord that is properly terminated in a grounding-type attachment plug having a fixed grounding contact member.

680.32 Ground-Fault Circuit Interrupters Required. All electrical equipment, including power-supply cords, used with storable pools shall be protected by ground-fault circuit interrupters.

680.33 Luminaires (Lighting Fixtures). An underwater luminaire (lighting fixture), if installed, shall be installed in or on the wall of the storable pool. It shall comply with one of the following two provisions.

(A) 15 Volts or Less. A luminaire (lighting fixture) shall be part of a cord-and-plug-connected lighting assembly. This assembly shall be listed as an assembly for the purpose and have the following construction features:

(1) No exposed metal parts
(2) A luminaire (fixture) lamp that operates at 15 volts or less

(3) An impact-resistant polymeric lens, luminaire (fixture) body, and transformer enclosure

(4) A transformer meeting the requirements of 680.23(A)(2) with a primary rating not over 150 volts

(B) Over 15 Volts But Not Over 150 Volts. A lighting assembly without a transformer and with the luminaire (fixture) lamp(s) operating at not over 150 volts shall be permitted to be cord-and-plug connected where the assembly is listed as an assembly for the purpose. The installation shall comply with 680.23(A)(5), and the assembly shall have the following construction features:

(1) No exposed metal parts

(2) An impact-resistant polymeric lens and luminaire (fixture) body

(3) A ground-fault circuit interrupter with open neutral protection as an integral part of the assembly

(4) The luminaire (fixture) lamp permanently connected to the ground-fault circuit interrupter with open-neutral protection

(5) Compliance with the requirements of 680.23(A)

IV. SPAS AND HOT TUBS

680.40 General. Electrical installations at spas and hot tubs shall comply with the provisions of Part I and Part IV of this article.

680.41 Emergency Switch for Spas and Hot Tubs. A clearly labeled emergency shutoff or control switch for the purpose of stopping the motor(s) that provide power to the recirculation system and jet system shall be installed at a point readily accessible to the users and not less than 1.5 m (5 ft) away, adjacent to, and within sight of the spa or hot tub. This requirement shall not apply to single-family dwellings.

680.42 Outdoor Installations. A spa or hot tub installed outdoors shall comply with the provisions of Parts I and II of this article, except as permitted in 680.42(A) and (B), that would otherwise apply to pools installed outdoors.

(A) Flexible Connections. Listed packaged spa or hot tub equipment assemblies or self-contained spas or hot tubs utilizing a factory-installed or assembled control panel or panelboard shall be permitted to use flexible connections as covered in 680.42(A)(1) and (A)(2).

(1) Flexible Conduit. Liquidtight flexible metal conduit or liquidtight flexible nonmetallic conduit shall be permitted in lengths of not more than 1.8 m (6 ft).

(2) Cord-and-Plug Connections. Cord-and-plug connections with a cord not longer than 4.6 m (15 ft) shall be permitted where protected by a ground-fault circuit interrupter.

(B) Bonding. Bonding by metal-to-metal mounting on a common frame or base shall be permitted. The metal bands or hoops used to secure wooden staves shall not be required to be bonded as required in 680.26.

(C) Interior Wiring to Outdoor Installations. In the interior of a one-family dwelling or in the interior of another building or structure associated with a one-family dwelling, any of the wiring methods recognized in Chapter 3 of this *Code* that contain a copper equipment grounding conductor that is insulated or enclosed within the outer sheath of the wiring method and not smaller than 12 AWG shall be permitted to be used for the connection to motor, heating, and control loads that are part of a self-contained spa or hot tub, or a packaged spa or hot tub equipment assembly. Wiring to an underwater light shall comply with 680.23 or 680.33.

680.43 Indoor Installations. A spa or hot tub installed indoors shall comply with the provisions of Parts I and II of this article except as modified by this section, and shall be connected by the wiring methods of Chapter 3.

Exception: Listed spa and hot tub packaged units rated 20 amperes or less shall be permitted to be cord-and-plug connected to facilitate the removal or disconnection of the unit for maintenance and repair.

(A) Receptacles. At least one 125-volt, 15- or 20-ampere receptacle on a general-purpose branch circuit shall be located not less than of 1.5 m (5 ft) from and not exceeding 3.0 m (10 ft) from the inside wall of the spa or hot tub.

(1) Location. Receptacles shall be located at least 1.5 m (5 ft) measured horizontally from the inside walls of the spa or hot tub.

(2) Protection, General. Receptacles rated 125 volts and 30 amperes or less and located within 3.0 m (10 ft) of the inside walls of a spa or hot tub shall be protected by a ground-fault circuit interrupter.

(3) Protection, Spa or Hot Tub Supply Receptacle. Receptacles that provide power for a spa or hot tub shall be ground-fault circuit-interrupter protected.

(4) Measurements. In determining the dimensions in this section addressing receptacle spacings, the distance to be measured shall be the shortest path the supply cord of an appliance connected to the receptacle would follow without piercing a floor, wall, ceiling, doorway with hinged or sliding door, window opening, or other effective permanent barrier.

(B) Installation of Luminaires (Lighting Fixtures), Lighting Outlets, and Ceiling-Suspended (Paddle) Fans.

(1) Elevation. Luminaires (lighting fixtures), except as covered in 680.43(B)(2), lighting outlets, and ceiling-suspended (paddle) fans located over the spa or hot tub or within 1.5 m (5 ft) from the inside walls of the spa or hot tub shall comply with the clearances specified in (a), (b), and (c) above the maximum water level.

    (a) Without GFCI. Where no GFCI protection is provided, the mounting height shall be not less than 3.7 m (12 ft).

    (b) With GFCI. Where GFCI protection is provided, the mounting height shall be permitted to be not less than 2.3 m (7 ft 6 in.).

    (c) Below 2.3 m (7 ft 6 in.). Luminaires (lighting fixtures) meeting the requirements of item (1) or (2) and protected by a ground-fault circuit interrupter shall be permitted to be installed less than 2.3 m (7 ft 6 in.) over a spa or hot tub.

(1) Recessed luminaires (fixtures) with a glass or plastic lens, nonmetallic or electrically isolated metal trim, and suitable for use in damp locations

(2) Surface-mounted luminaires (fixtures) with a glass or plastic globe, a nonmetallic body, or a metallic body isolated from contact, and suitable for use in damp locations

(2) Underwater Applications. Underwater luminaires (lighting fixtures) shall comply with the provisions of 680.23 or 680.33.

(C) Wall Switches. Switches shall be located at least 1.5 m (5 ft), measured horizontally, from the inside walls of the spa or hot tub.

(D) Bonding. The following parts shall be bonded together:

(1) All metal fittings within or attached to the spa or hot tub structure
(2) Metal parts of electrical equipment associated with the spa or hot tub water circulating system, including pump motors
(3) Metal conduit and metal piping that are within 1.5 m (5 ft) of the inside walls of the spa or hot tub and that are not separated from the spa or hot tub by a permanent barrier
(4) All metal surfaces that are within 1.5 m (5 ft) of the inside walls of the spa or hot tub and that are not separated from the spa or hot tub area by a permanent barrier

Exception: Small conductive surfaces not likely to become energized, such as air and water jets and drain fittings, where not connected to metallic piping, towel bars, mirror frames, and similar nonelectrical equipment, shall not be required to be bonded.

(5) Electrical devices and controls that are not associated with the spas or hot tubs and that are located not less than 1.5 m (5 ft) from such units; otherwise they shall be bonded to the spa or hot tub system

(E) Methods of Bonding. All metal parts associated with the spa or hot tub shall be bonded by any of the following methods:

(1) The interconnection of threaded metal piping and fittings
(2) Metal-to-metal mounting on a common frame or base
(3) The provisions of a copper bonding jumper, insulated, covered, or bare, not smaller than 8 AWG solid

(F) Grounding. The following equipment shall be grounded:

(1) All electric equipment located within 1.5 m (5 ft) of the inside wall of the spa or hot tub
(2) All electric equipment associated with the circulating system of the spa or hot tub

680.44 Protection. Except as otherwise provided in this section, the outlet(s) that supplies a self-contained spa or hot tub, a packaged spa or hot tub equipment assembly, or a field-assembled spa or hot tub shall be protected by a ground-fault circuit interrupter.

(A) Listed Units. If so marked, a listed self-contained unit or listed packaged equipment assembly that includes integral ground-fault circuit-interrupter protection for all electrical parts within the unit or assembly (pumps, air blowers, heaters, lights, controls, sanitizer generators, wiring, and so forth) shall be permitted without additional GFCI protection.

(B) Other Units. A field assembled spa or hot tub rated 3 phase or rated over 250 volts or with a heater load of more than 50 amperes shall not require the supply to be protected by a ground-fault circuit interrupter.

(C) Combination Pool and Spa or Hot Tub. A combination pool/hot tub or spa assembly commonly bonded need not be protected by a ground-fault circuit interrupter.

V. FOUNTAINS

680.50 General. The provisions of Part I and Part V of this article shall apply to all permanently installed fountains as defined in 680.2. Fountains that have water common to a pool shall additionally comply with the requirements in Part II of this article. Part V does not cover self-contained, portable fountains not larger than 1.5 m (5 ft) in any dimension. Portable fountains shall comply with Parts II and III of Article 422.

680.51 Luminaires (Lighting Fixtures), Submersible Pumps, and Other Submersible Equipment.

(A) Ground-Fault Circuit Interrupter. Fountain equipment, unless listed for operation at 15 volts or less and supplied by a transformer that complies with 680.23(A)(2), shall be protected by a ground-fault circuit interrupter.

(B) Operating Voltage. No luminaires (lighting fixtures) shall be installed for operation on supply circuits over 150 volts between conductors. Submersible pumps and other submersible equipment shall operate at 300 volts or less between conductors.

(C) Luminaire (Lighting Fixture) Lenses. Luminaires (lighting fixtures) shall be installed with the top of the luminaire (fixture) lens below the normal water level of the fountain unless listed for above-water locations. A luminaire (lighting fixture) facing upward shall have the lens adequately guarded to prevent contact by any person.

(D) Overheating Protection. Electrical equipment that depends on submersion for safe operation shall be protected against overheating by a low-water cutoff or other approved means when not submerged.

(E) Wiring. Equipment shall be equipped with provisions for threaded conduit entries or be provided with a suitable flexible cord. The maximum length of exposed cord in the fountain shall be limited to 3.0 m (10 ft). Cords extending beyond the fountain perimeter shall be enclosed in approved wiring enclosures. Metal parts of equipment in contact with water shall be of brass or other approved corrosion-resistant metal.

(F) Servicing. All equipment shall be removable from the water for relamping or normal maintenance. Luminaires (fixtures) shall not be permanently embedded into the fountain structure such that the water level must be reduced or the fountain drained for relamping, maintenance, or inspection.

(G) Stability. Equipment shall be inherently stable or be securely fastened in place.

680.52 Junction Boxes and Other Enclosures.

(A) General. Junction boxes and other enclosures used for other than underwater installation shall comply with 680.24.

(B) Underwater Junction Boxes and Other Underwater Enclosures. Junction boxes and other underwater enclosures shall meet the requirements of 680.52(B)(1) and (B)(2).

(1) Construction.

   (a) Underwater enclosures shall be equipped with provisions for threaded conduit entries or compression glands or seals for cord entry.

(b) Underwater enclosures shall be submersible, and made of copper, brass, or other approved corrosion-resistant material.

(2) Installation. Underwater enclosure installations shall comply with (a) and (b).

(a) Underwater enclosures shall be filled with an approved potting compound to prevent the entry of moisture.

(b) Underwater enclosures shall be firmly attached to the supports or directly to the fountain surface and bonded as required. Where the junction box is supported only by the conduit, the conduit shall be of copper, brass, or other approved corrosion-resistant metal. Where the box is fed by nonmetallic conduit, it shall have additional supports and fasteners of copper, brass, or other approved corrosion-resistant material.

680.53 Bonding. All metal piping systems associated with the fountain shall be bonded to the equipment grounding conductor of the branch circuit supplying the fountain.

680.54 Grounding. The following equipment shall be grounded:

(1) All electrical equipment located within the fountain or within 1.5 m (5 ft) of the inside wall of the fountain
(2) All electrical equipment associated with the recirculating system of the fountain
(3) Panelboards that are not part of the service equipment and that supply any electrical equipment associated with the fountain

680.55 Methods of Grounding.

(A) Applied Provisions. The provisions of 680.21(A), 680.23(B)(3), 680.23(F)(1) and (2), 680.24(F), and 680.25 shall apply.

(B) Supplied by a Flexible Cord. Electrical equipment that is supplied by a flexible cord shall have all exposed non–current-carrying metal parts grounded by an insulated copper equipment grounding conductor that is an integral part of this cord. The grounding conductor shall be connected to a grounding terminal in the supply junction box, transformer enclosure, or other enclosure.

680.56 Cord-and-Plug-Connected Equipment.

(A) Ground-Fault Circuit Interrupter. All electrical equipment, including power-supply cords, shall be protected by ground-fault circuit interrupters.

(B) Cord Type. Flexible cord immersed in or exposed to water shall be of a type for extra-hard usage, as designated in Table 400.4 and shall be a listed type with a "W" suffix.

(C) Sealing. The end of the flexible cord jacket and the flexible cord conductor termination within equipment shall be covered with, or encapsulated in, a suitable potting compound to prevent the entry of water into the equipment through the cord or its conductors. In addition, the ground connection within equipment shall be similarly treated to protect such connections from the deteriorating effect of water that may enter into the equipment.

(D) Terminations. Connections with flexible cord shall be permanent, except that grounding-type attachment plugs and receptacles shall be permitted to facilitate removal or disconnection for maintenance, repair, or storage of fixed or stationary equipment not located in any water-containing part of a fountain.

VII. HYDROMASSAGE BATHTUBS

680.70 General. Hydromassage bathtubs as defined in 680.2 shall comply with Part VII of this article. They shall not be required to comply with other parts of this article.

680.71 Protection. Hydromassage bathtubs and their associated electrical components shall be protected by a ground-fault circuit interrupter. All 125-volt, single-phase receptacles not exceeding 30 amperes and located within 1.5 m (5 ft) measured horizontally of the inside walls of a hydromassage tub shall be protected by a ground-fault circuit interrupter(s).

680.72 Other Electrical Equipment. Luminaires (lighting fixtures), switches, receptacles, and other electrical equipment located in the same room, and not directly associated with a hydromassage bathtub, shall be installed in accordance with the requirements of Chapters 1 through 4 in this *Code* covering the installation of that equipment in bathrooms.

680.73 Accessibility. Hydromassage bathtub electrical equipment shall be accessible without damaging the building structure or building finish.

680.74 Bonding. All metal piping systems, metal parts of electrical equipment, and pump motors associated with the hydromassage tub shall be bonded together using a copper bonding jumper, insulated, covered, or bare, not smaller than 8 AWG solid. Metal parts of listed equipment incorporating an approved system of double insulation and providing a means for grounding internal nonaccessible, non–current-carrying metal parts shall not be bonded.

CHAPTER 26

EXISTING DWELLINGS

INTRODUCTION

This chapter contains provisions from one section of Article 406—Receptacles, Cord Connectors, and Attachment Plugs (Caps). Covered in this chapter are the specifications for replacing receptacles. This sections' requirements cover grounding-type receptacles, ground-fault circuit-interrupters, and nongrounding-type receptacles. Provisions for installing new receptacles are in Articles 210 and 406 (see Chapter 17 of this book).

ARTICLE 406
Receptacles, Cord Connectors, and Attachment Plugs (Caps)

406.3 General Installation Requirements.

(D) Replacements. Replacement of receptacles shall comply with 406.3(D)(1), (2), and (3) as applicable.

(1) Grounding-Type Receptacles. Where a grounding means exists in the receptacle enclosure or a grounding conductor is installed in accordance with 250.130(C), grounding-type receptacles shall be used and shall be connected to the grounding conductor in accordance with 406.3(C) or 250.130(C).

(2) Ground-Fault Circuit Interrupters. Ground-fault circuit-interrupter protected receptacles shall be provided where replacements are made at receptacle outlets that are required to be so protected elsewhere in this *Code*.

(3) Nongrounding-Type Receptacles. Where grounding means does not exist in the receptacle enclosure, the installation shall comply with (a), (b), or (c).

(a) A nongrounding-type receptacle(s) shall be permitted to be replaced with another nongrounding-type receptacle(s).

(b) A nongrounding-type receptacle(s) shall be permitted to be replaced with a ground-fault circuit interrupter-type of receptacle(s). These receptacles shall be marked "No Equipment Ground." An equipment grounding conductor shall not be connected from the ground-fault circuit interrupter-type receptacle to any outlet supplied from the ground-fault circuit-interrupter receptacle.

(c) A nongrounding-type receptacle(s) shall be permitted to be replaced with a grounding-type receptacle(s) where supplied through a ground-fault circuit interrupter. Grounding-type receptacles supplied through the ground-fault circuit interrupter shall be marked "GFCI Protected" and "No Equipment Ground." An equipment grounding conductor shall not be connected between the grounding-type receptacles.

CHAPTER 27

CONDUIT AND TUBING FILL TABLES

INTRODUCTION

This chapter contains tables and notes from Chapter 9—Tables and Annex C—Conduit and Tubing Fill Tables for Conductors and Fixture Wires of the Same Size. These tables are used to determine the maximum number of conductors permitted in raceways (conduit and tubing). Tables 4 and 5 of Chapter 9 are used when different size conductors are installed in the same conduit, and Annex C is used when all the conductors are the same size. Table 8—Conductor Properties (in Chapter 9) provides useful information such as size (AWG or kcmil), circular mils, overall area, and the resistance of copper and aluminum conductors.

CHAPTER 9 TABLES

Notes to Tables

(1) See Annex C for the maximum number of conductors and fixture wires, all of the same size (total cross-sectional area including insulation) permitted in trade sizes of the applicable conduit or tubing.

(2) Table 1 applies only to complete conduit or tubing systems and is not intended to apply to sections of conduit or tubing used to protect exposed wiring from physical damage.

(3) Equipment grounding or bonding conductors, where installed, shall be included when calculating conduit or tubing fill. The actual dimensions of the equipment grounding or bonding conductor (insulated or bare) shall be used in the calculation.

(4) Where conduit or tubing nipples having a maximum length not to exceed 600 mm (24 in.) are installed between boxes, cabinets, and similar enclosures, the nipples shall be permitted to be filled to 60 percent of their total cross-sectional area, and 310.15(B)(2)(a) adjustment factors need not apply to this condition.

Table 1 Percent of Cross Section of Conduit and Tubing for Conductors

Number of Conductors	All Conductor Types
1	53
2	31
Over 2	40

FPN No. 1: Table 1 is based on common conditions of proper cabling and alignment of conductors where the length of the pull and the number of bends are within reasonable limits. It should be recognized that, for certain conditions, a larger size conduit or a lesser conduit fill should be considered.

FPN No. 2: When pulling three conductors or cables into a raceway, if the ratio of the raceway (inside diameter) to the conductor or cable (outside diameter) is between 2.8 and 3.2, jamming can occur. While jamming can occur when pulling four or more conductors or cables into a raceway, the probability is very low.

(5) For conductors not included in Chapter 9, such as multiconductor cables, the actual dimensions shall be used.

(6) For combinations of conductors of different sizes, use Table 5 and Table 5A for dimensions of conductors and Table 4 for the applicable conduit or tubing dimensions.

(7) When calculating the maximum number of conductors permitted in a conduit or tubing, all of the same size (total cross-sectional area including insulation), the next higher whole number shall be used to determine the maximum number of conductors permitted when the calculation results in a decimal of 0.8 or larger.

(8) Where bare conductors are permitted by other sections of this *Code,* the dimensions for bare conductors in Table 8 shall be permitted.

(9) A multiconductor cable of two or more conductors shall be treated as a single conductor for calculating percentage conduit fill area. For cables that have elliptical cross sections, the cross-sectional area calculation shall be based on using the major diameter of the ellipse as a circle diameter.

Table 4 Dimensions and Percent Area of Conduit and Tubing
(Areas of Conduit or Tubing for the Combinations of Wires Permitted in Table 1)

Metric Designator	Trade Size	Nominal Internal Diameter mm	Nominal Internal Diameter in.	Total Area 100% mm²	Total Area 100% in.²	2 Wires 31% mm²	2 Wires 31% in.²	Over 2 Wires 40% mm²	Over 2 Wires 40% in.²	1 Wire 53% mm²	1 Wire 53% in.²	60% mm²	60% in.²
						Article 358—Electrical Metallic Tubing (EMT)							
16	1/2	15.8	0.622	196	0.304	61	0.094	78	0.122	104	0.161	118	0.182
21	3/4	20.9	0.824	343	0.533	106	0.165	137	0.213	182	0.283	206	0.320
27	1	26.6	1.049	556	0.864	172	0.268	222	0.346	295	0.458	333	0.519
35	1-1/4	35.1	1.380	968	1.496	300	0.464	387	0.598	513	0.793	581	0.897
41	1-1/2	40.9	1.610	1314	2.036	407	0.631	526	0.814	696	1.079	788	1.221
53	2	52.5	2.067	2165	3.356	671	1.040	866	1.342	1147	1.778	1299	2.013
63	2-1/2	69.4	2.731	3783	5.858	1173	1.816	1513	2.343	2005	3.105	2270	3.515
78	3	85.2	3.356	5701	8.846	1767	2.742	2280	3.538	3022	4.688	3421	5.307
91	3-1/2	97.4	3.834	7451	11.545	2310	3.579	2980	4.618	3949	6.119	4471	6.927
103	4	110.1	4.334	9521	14.753	2951	4.573	3808	5.901	5046	7.819	5712	8.852

Article 362—Electrical Nonmetallic Tubing (ENT)

16	1/2	14.2	0.560	158	0.246	49	0.076	63	0.099	84	0.131	95	0.148
21	3/4	19.3	0.760	293	0.454	91	0.141	117	0.181	155	0.240	176	0.272
27	1	25.4	1.000	507	0.785	157	0.243	203	0.314	269	0.416	304	0.471
35	1-1/4	34.0	1.340	908	1.410	281	0.437	363	0.564	481	0.747	545	0.846
41	1-1/2	39.9	1.570	1250	1.936	388	0.600	500	0.774	663	1.026	750	1.162
53	2	51.3	2.020	2067	3.205	641	0.993	827	1.282	1095	1.699	1240	1.923
63	2-1/2	—	—	—	—	—	—	—	—	—	—	—	—
78	3	—	—	—	—	—	—	—	—	—	—	—	—
91	3-1/2	—	—	—	—	—	—	—	—	—	—	—	—

Article 348—Flexible Metal Conduit (FMC)

12	3/8	9.7	0.384	74	0.116	23	0.036	30	0.046	39	0.061	44	0.069
16	1/2	16.1	0.635	204	0.317	63	0.098	81	0.127	108	0.168	122	0.190
21	3/4	20.9	0.824	343	0.533	106	0.165	137	0.213	182	0.283	206	0.320
27	1	25.9	1.020	527	0.817	163	0.253	211	0.327	279	0.433	316	0.490
35	1-1/4	32.4	1.275	824	1.277	256	0.396	330	0.511	437	0.677	495	0.766
41	1-1/2	39.1	1.538	1201	1.858	372	0.576	480	0.743	636	0.985	720	1.115
53	2	51.8	2.040	2107	3.269	653	1.013	843	1.307	1117	1.732	1264	1.961

Continued

Table 4 Dimensions and Percent Area of Conduit and Tubing
(Areas of Conduit or Tubing for the Combinations of Wires Permitted in Table 1) *Continued*

Metric. Designator	Trade Size	Nominal Internal Diameter		Total Area 100%		2 Wires 31%		Over 2 Wires 40%		1 Wire 53%		60%	
		mm	in.	mm²	in.²	mm²	in.²	mm²	in.²	mm²	in.²	mm²	in.²
Article 348—Flexible Metal Conduit (FMT) *Continued*													
63	2-1/2	63.5	2.500	3167	4.909	982	1.522	1267	1.963	1678	2.602	1900	2.945
78	3	76.2	3.000	4560	7.069	1414	2.191	1824	2.827	2417	3.746	2736	4.241
91	3-1/2	88.9	3.500	6207	9.621	1924	2.983	2483	3.848	3290	5.099	3724	5.773
103	4	101.6	4.000	8107	12.566	2513	3.896	3243	5.027	4297	6.660	4864	7.540
Article 342—Intermediate Metal Conduit (IMC)													
12	3/8	—	—	—	—	—	—	—	—	—	—	—	—
16	1/2	16.8	0.660	222	0.342	69	0.106	89	0.137	117	0.181	133	0.205
21	3/4	21.9	0.864	377	0.586	117	0.182	151	0.235	200	0.311	226	0.352
27	1	28.1	1.105	620	0.959	192	0.297	248	0.384	329	0.508	372	0.575
35	1-1/4	36.8	1.448	1064	1.647	330	0.510	425	0.659	564	0.873	638	0.988
41	1-1/2	42.7	1.683	1432	2.225	444	0.690	573	0.890	759	1.179	859	1.335

53	2	54.6	2.150	2341	3.630	726	1.125	937	1.452	1241	1.924	1405	2.178
63	2-1/2	64.9	2.557	3308	5.135	1026	1.592	1323	2.054	1753	2.722	1985	3.081
78	3	80.7	3.176	5115	7.922	1586	2.456	2046	3.169	2711	4.199	3069	4.753
91	3-1/2	93.2	3.671	6822	10.584	2115	3.281	2729	4.234	3616	5.610	4093	6.351
103	4	105.4	4.166	8725	13.631	2705	4.226	3490	5.452	4624	7.224	5235	8.179

Article 356—Liquidtight Flexible Nonmetallic Conduit (LFNC-B*)

12	3/8	12.5	0.494	123	0.192	38	0.059	49	0.077	65	0.102	74	0.115
16	1/2	16.1	0.632	204	0.314	63	0.097	81	0.125	108	0.166	122	0.187
21	3/4	21.1	0.830	350	0.541	108	0.168	140	0.216	185	0.287	210	0.325
27	1	26.8	1.054	564	0.873	175	0.270	226	0.349	299	0.462	338	0.524
35	1-1/4	35.4	1.395	984	1.528	305	0.474	394	0.611	522	0.810	591	0.917
41	1-1/2	40.3	1.588	1276	1.981	395	0.614	510	0.792	676	1.050	765	1.188
53	2	51.6	2.033	2091	3.246	648	1.006	836	1.298	1108	1.720	1255	1.948

Article 356—Liquidtight Flexible Nonmetallic Conduit (LFNC-A*)

12	3/8	12.6	0.495	125	0.192	39	0.060	50	0.077	66	0.102	75	0.115
16	1/2	16.0	0.630	201	0.312	62	0.097	80	0.125	107	0.165	121	0.187
21	3/4	21.0	0.825	346	0.535	107	0.166	139	0.214	184	0.283	208	0.321

Continued

Table 4 Dimensions and Percent Area of Conduit and Tubing
(Areas of Conduit or Tubing for the Combinations of Wires Permitted in Table 1) *Continued*

Metric Designator	Trade Size	Nominal Internal Diameter		Total Area 100%		2 Wires 31%		Over 2 Wires 40%		1 Wire 53%		60%	
		mm	in.	mm²	in.²	mm²	in.²	mm²	in.²	mm²	in.²	mm²	in.²
Article 356—Liquidtight Flexible Nonmetallic Conduit (LFNC-A*) *Continued*													
27	1	26.5	1.043	552	0.854	171	0.265	221	0.342	292	0.453	331	0.513
35	1-1/4	35.1	1.383	968	1.502	300	0.466	387	0.601	513	0.796	581	0.901
41	1-1/2	40.7	1.603	1301	2.018	403	0.626	520	0.807	690	1.070	781	1.211
53	2	52.4	2.063	2157	3.343	669	1.036	863	1.337	1143	1.772	1294	2.006
Article 350—Liquidtight Flexible Metal Conduit (LFMC)													
12	1/8	12.5	0.494	123	0.192	38	0.059	49	0.077	65	0.102	74	0.115
16	1/2	16.1	0.632	204	0.314	63	0.097	81	0.125	108	0.166	122	0.188
21	1/4	21.1	0.830	350	0.541	108	0.168	140	0.216	185	0.287	210	0.325
27	1	26.8	1.054	564	0.873	175	0.270	226	0.349	299	0.462	338	0.524
35	1-1/4	35.4	1.395	984	1.528	305	0.474	394	0.611	522	0.810	591	0.917
41	1-1/2	40.3	1.588	1276	1.981	395	0.614	510	0.792	676	1.050	765	1.188

53	2	51.6	2.033	2091	3.246	648	1.006	836	1.298	1108	1.720	1255	1.948
63	2-1/2	63.3	2.493	3147	4.881	976	1.513	1259	1.953	1668	2.587	1888	2.929
78	3	78.4	3.085	4827	7.475	1497	2.317	1931	2.990	2559	3.962	2896	4.485
91	3-1/2	89.4	3.520	6277	9.731	1946	3.017	2511	3.893	3327	5.158	3766	5.839
103	4	102.1	4.020	8187	12.692	2538	3.935	3275	5.077	4339	6.727	4912	7.615
129	5	—	—	—	—	—	—	—	—	—	—	—	—
155	6	—	—	—	—	—	—	—	—	—	—	—	—

Article 344—Rigid Metal Conduit

12	3/8	—	—	—	—	—	—	—	—	—	—	—	—
16	1/2	16.1	0.632	203	0.314	63	0.097	81	0.125	108	0.166	122	0.188
21	3/4	21.2	0.836	353	0.549	109	0.170	141	0.220	187	0.291	212	0.329
27	1	27.0	1.063	573	0.887	177	0.275	229	0.355	303	0.470	344	0.532
35	1-1/4	35.4	1.394	984	1.526	305	0.473	394	0.610	522	0.809	591	0.916
41	1-1/2	41.2	1.624	1333	2.071	413	0.642	533	0.829	707	1.098	800	1.243
53	2	52.9	2.083	2198	3.408	681	1.056	879	1.363	1165	1.806	1319	2.045
63	2-1/2	63.2	2.489	3137	4.866	972	1.508	1255	1.946	1663	2.579	1882	2.919
78	3	78.5	3.090	4840	7.499	1500	2.325	1936	3.000	2565	3.974	2904	4.499
91	3-1/2	90.7	3.570	6461	10.010	2003	3.103	2584	4.004	3424	5.305	3877	6.006
103	4	102.9	4.050	8316	12.882	2578	3.994	3326	5.153	4408	6.828	4990	7.729

Continued

Table 4 Dimensions and Percent Area of Conduit and Tubing
(Areas of Conduit or Tubing for the Combinations of Wires Permitted in Table 1) *Continued*

Metric Designator	Trade Size	Nominal Internal Diameter		Total Area 100%		2 Wires 31%		Over 2 Wires 40%		1 Wire 53%		60%	
		mm	in.	mm²	in.²	mm²	in.²	mm²	in.²	mm²	in.²	mm²	in.²
						Article 344—Rigid Metal Conduit *Continued*							
129	5	128.9	5.073	13050	20.212	4045	6.266	5220	8.085	6916	10.713	7830	12.127
155	6	154.8	6.093	18812	29.158	5834	9.039	7528	11.663	9975	15.454	11292	17.495
						Article 352—Rigid PVC Conduit (RNC), Schedule 80							
12	1/8	—	—	—	—	—	—	—	—	—	—	—	—
16	1/2	13.4	0.526	141	0.217	44	0.067	56	0.087	75	0.115	85	0.130
21	3/4	18.3	0.722	263	0.409	82	0.127	105	0.164	139	0.217	158	0.246
27	1	23.8	0.936	445	0.688	138	0.213	178	0.275	236	0.365	267	0.413
35	1-1/4	31.9	1.255	799	1.237	248	0.383	320	0.495	424	0.656	480	0.742
41	1-1/2	37.5	1.476	1104	1.711	342	0.530	442	0.684	585	0.907	663	1.027
53	2	48.6	1.913	1855	2.874	575	0.891	742	1.150	983	1.523	1113	1.725
63	2-1/2	58.2	2.290	2660	4.119	825	1.277	1064	1.647	1410	2.183	1596	2.471

78	3	72.7	2.864	4151	6.442	1287	1.997	1660	2.577	2200	3.414	2491	3.865
91	3-1/2	84.5	3.326	5608	8.688	1738	2.693	2243	3.475	2972	4.605	3365	5.213
103	4	96.2	3.786	7268	11.258	2253	3.490	2907	4.503	3852	5.967	4361	6.755
129	5	121.1	4.768	11518	17.855	3571	5.535	4607	7.142	6105	9.463	6911	10.713
155	6	145.0	5.709	16513	25.598	5119	7.935	6605	10.239	8752	13.567	9908	15.359

Article 352—Rigid PVC Conduit (RNC), Schedule 40, and HDPE Conduit

12	1/8	—	—	—	—	—	—	—	—	—	—	—	—
16	1/2	15.3	0.602	184	0.285	57	0.088	74	0.114	97	0.151	110	0.171
21	3/4	20.4	0.804	327	0.508	101	0.157	131	0.203	173	0.269	196	0.305
27	1	26.1	1.029	535	0.832	166	0.258	214	0.333	284	0.441	321	0.499
35	1-1/4	34.5	1.360	935	1.453	290	0.450	374	0.581	495	0.770	561	0.872
41	1-1/2	40.4	1.590	1282	1.986	397	0.616	513	0.794	679	1.052	769	1.191
53	2	52.0	2.047	2124	3.291	658	1.020	849	1.316	1126	1.744	1274	1.975
63	2-1/2	62.1	2.445	3029	4.695	939	1.455	1212	1.878	1605	2.488	1817	2.817
78	3	77.3	3.042	4693	7.268	1455	2.253	1877	2.907	2487	3.852	2816	4.361
91	3-1/2	89.4	3.521	6277	9.737	1946	3.018	2511	3.895	3327	5.161	3766	5.842
103	4	101.5	3.998	8091	12.554	2508	3.892	3237	5.022	4288	6.654	4855	7.532
129	5	127.4	5.016	12748	19.761	3952	6.126	5099	7.904	6756	10.473	7649	11.856
155	6	153.2	6.031	18433	28.567	5714	8.856	7373	11.427	9770	15.141	11060	17.140

Table 5   Dimensions of Insulated Conductors and Fixture Wires

Type: FFH-2, RFH-1, RFH-2, RHH*, RHW*, RHW-2*, RHH, RHW, RHW-2, SF-1, SF-2, SFF-1, SFF-2, TF, TFF, THHW, THW, THW-2, TW, XF, XFF

Type	Size (AWG or kcmil)	Approximate Diameter		Approximate Area	
		mm	in.	mm²	in.²
RFH-2, FFH-2	18	3.454	0.136	9.355	0.0145
	16	3.759	0.148	11.10	0.0172
RHW-2, RHH, RHW	14	4.902	0.193	18.90	0.0293
	12	5.385	0.212	22.77	0.0353
	10	5.994	0.236	28.19	0.0437
	8	8.280	0.326	53.87	0.0835
	6	9.246	0.364	67.16	0.1041
	4	10.46	0.412	86.00	0.1333
	3	11.18	0.440	98.13	0.1521
	2	11.99	0.472	112.9	0.1750
	1	14.78	0.582	171.6	0.2660
	1/0	15.80	0.622	196.1	0.3039
	2/0	16.97	0.668	226.1	0.3505
	3/0	18.29	0.720	262.7	0.4072
	4/0	19.76	0.778	306.7	0.4754

250	22.73	0.895	405.9	0.6291
300	24.13	0.950	457.3	0.7088
350	25.43	1.001	507.7	0.7870
400	26.62	1.048	556.5	0.8626
500	28.78	1.133	650.5	1.0082
600	31.57	1.243	782.9	1.2135
700	33.38	1.314	874.9	1.3561
750	34.24	1.348	920.8	1.4272
800	35.05	1.380	965.0	1.4957
900	36.68	1.444	1057	1.6377
1000	38.15	1.502	1143	1.7719
1250	43.92	1.729	1515	2.3479
1500	47.04	1.852	1738	2.6938
1750	49.94	1.966	1959	3.0357
2000	52.63	2.072	2175	3.3719

SF-2, SFF-2

18	3.073	0.121	7.419	0.0115
16	3.378	0.133	8.968	0.0139
14	3.759	0.148	11.10	0.0172

Continued

Table 5 Dimensions of Insulated Conductors and Fixture Wires *Continued*

Type	Size (AWG or kcmil)	Approximate Diameter		Approximate Area	
		mm	in.	mm²	in.²
Type: FFH-2, RFH-1, RFH-2, RHH*, RHW*, RHW-2*, RHH, RHW, RHW-2, SF-1, SF-2, SFF-1, SFF-2, TF, TFF, THHW, THW, THW-2, TW, XF, XFF *Continued*					
SF-1, SFF-1	18	2.311	0.091	4.194	0.0065
RFH-1, XF, XFF	18	2.692	0.106	5.161	0.0080
TF, TFF, XF, XFF	16	2.997	0.118	7.032	0.0109
TW, XF, XFF, THHW, THW, THW-2	14	3.378	0.133	8.968	0.0139
TW, THHW, THW, THW-2	12	3.861	0.152	11.68	0.0181
	10	4.470	0.176	15.68	0.0243
	8	5.994	0.236	28.19	0.0437
RHH*, RHW*, RHW-2*	14	4.140	0.163	13.48	0.0209
	12	4.623	0.182	16.77	0.0260

Type: RHH*, RHW*, RHW-2*, THHN, THHW, THW, THW-2, TFN, TFFN, THWN, THWN-2, XF, XFF

	Size				
THHW, THW, AF, XF, XFF	10	5.232	0.206	21.48	0.0333
RHH*, RHW*, RHW-2*	8	6.756	0.266	35.87	0.0556
TW, THW, THHW, THW-2, RHH*, RHW*, RHW-2*	6	7.722	0.304	46.84	0.0726
	4	8.941	0.652	62.77	0.0973
	3	9.652	0.380	73.16	0.1134
	2	10.46	0.412	86.00	0.1333
	1	12.50	0.492	122.6	0.1901
	1/0	13.51	0.532	143.4	0.2223
	2/0	14.68	0.578	169.3	0.2624
	3/0	16.00	0.630	201.1	0.3117
	4/0	17.48	0.688	239.9	0.3718
	250	19.43	0.765	296.5	0.4596
	300	20.93	0.820	340.7	0.5281
	350	22.12	0.871	384.4	0.5958
	400	23.32	0.918	427.0	0.6619

Continued

Table 5 Dimensions of Insulated Conductors and Fixture Wires *Continued*

Type	Size (AWG or kcmil)	Approximate Diameter		Approximate Area	
		mm	in.	mm^2	in.^{2}
Type: RHH*, RHW*, RHW-2*, THHN, THHW, THW, THW-2, TFN, TFFN, THWN, THWN-2, XF, XFF *Continued*					
TW, THW, THHW, THW-2, RHH*, RHW*, RHW-2*	500	25.48	1.003	509.7	0.7901
	600	28.27	1.113	627.7	0.9729
	700	30.07	1.184	710.3	1.1010
	750	30.94	1.218	751.7	1.1652
	800	31.75	1.250	791.7	1.2272
	900	33.38	1.314	874.9	1.3561
	1000	34.85	1.372	953.8	1.4784
	1250	39.09	1.539	1200	1.8602
	1500	42.21	1.662	1400	2.1695
	1750	45.11	1.776	1598	2.4773
	2000	47.80	1.882	1795	2.7818

TFN, TFFN					
18	2.134	0.084		3.548	0.0055
16	2.438	0.096		4.645	0.0072
THHN, THWN, THWN-2					
14	2.819	0.111		6.258	0.0097
12	3.302	0.130		8.581	0.0133
10	4.166	0.164		13.61	0.0211
8	5.486	0.216		23.61	0.0366
6	6.452	0.254		32.71	0.0507
4	8.230	0.324		53.16	0.0824
3	8.941	0.352		62.77	0.0973
2	9.754	0.384		74.71	0.1158
1	11.33	0.446		100.8	0.1562
1/0	12.34	0.486		119.7	0.1855
2/0	13.51	0.532		143.4	0.2223
3/0	14.83	0.584		172.8	0.2679
4/0	16.31	0.642		208.8	0.3237
250	18.06	0.711		256.1	0.3970
300	19.46	0.766		297.3	0.4608

Continued

Table 5 Dimensions of Insulated Conductors and Fixture Wires *Continued*

Type	Size (AWG or kcmil)	Approximate Diameter		Approximate Area	
		mm	in.	mm²	in.²
Type: FEP, FEPB, PAF, PAFF, PF PFA, PFAH, PFF, PGF, PGFF, PTF, PTFE, TFE, THHN, THWN, THWN-2, Z, ZF, ZFF					
THHN, THWN, THWN-2	350	20.75	0.817	338.2	0.5242
	400	21.95	0.864	378.3	0.5863
	500	24.10	0.949	456.3	0.7073
	600	26.70	1.051	559.7	0.8676
	700	28.50	1.122	637.9	0.9887
	750	29.36	1.156	677.2	1.0496
	800	30.18	1.188	715.2	1.1085
	900	31.80	1.252	794.3	1.2311
	1000	33.27	1.310*	869.5	1.3478
PF, PGFF, PGF, PFF, PTF, PAF, PTFF, PAFF	18	2.184	0.086	3.742	0.0058
	16	2.489	0.098	4.839	0.0075
PF, PGFF, PGF, PFF, PTF, PAF, PTFF, PAFF, TFE, FEP, PFA, FEPB, PFAH	14	2.870	0.113	6.452	0.0100

TFE, FEP, PFA, FEPB, PFAH	12	3.353	0.132	8.839	0.0137
	10	3.962	0.156	12.32	0.0191
	8	5.232	0.206	21.48	0.0333
	6	6.198	0.244	30.19	0.0468
	4	7.417	0.292	43.23	0.0670
	3	8.128	0.320	51.87	0.0804
	2	8.941	0.352	62.77	0.0973
TFE, PFAH	1	10.72	0.422	90.26	0.1399
TFE, PFA, PFAH, Z	1/0	11.73	0.462	108.1	0.1676
	2/0	12.90	0.508	130.8	0.2027
	3/0	14.22	0.560	158.9	0.2463
	4/0	15.70	0.618	193.5	0.3000
ZF, ZFF	18	1.930	0.076	2.903	0.0045
	16	2.235	0.088	3.935	0.0061
Z, ZF, ZFF	14	2.616	0.103	5.355	0.0083
Z	12	3.099	0.122	7.548	0.0117
	10	3.962	0.156	12.32	0.0191

Continued

Table 5 Dimensions of Insulated Conductors and Fixture Wires *Continued*

Type	Size (AWG or kcmil)	Approximate Diameter		Approximate Area		
		mm	in.	mm²	in.²	

Type: FEP, FEPB, PAF, PAFF, PF PFA, PFAH, PF PGF, PGFF, PTF, PTFE, TFE, THHN, THWN, THWN-2, Z, ZF, ZFF *Continued*

Type	Size	mm	in.	mm²	in.²	
Z	8	4.978	0.196	19.48	0.0302	
	6	5.944	0.234	27.74	0.0430	
	4	7.163	0.282	40.32	0.0625	
	3	8.382	0.330	55.16	0.0855	
	2	9.195	0.362	66.39	0.1029	
	1	10.21	0.402	81.87	0.1269	

Type: KF-1, KF-2, KFF-1, KFF-2, XHH, XHHW, XHHW-2, ZW

Type	Size	mm	in.	mm²	in.²	
XHHW, ZW, XHHW-2, XHH	14	3.378	0.133	8.968	0.0139	
	12	3.861	0.152	11.68	0.0181	
	10	4.470	0.176	15.68	0.0243	
	8	5.994	0.236	28.19	0.0437	
	6	6.960	0.274	38.06	0.0590	
	4	8.179	0.322	52.52	0.0814	
	3	8.890	0.350	62.06	0.0962	
	2	9.703	0.382	73.94	0.1146	

XHHW, XHHW-2, XHH

1	11.23	0.442	98.97	0.1534
1/0	12.24	0.482	117.7	0.1825
2/0	13.41	0.528	141.3	0.2190
3/0	14.73	0.580	170.5	0.2642
4/0	16.21	0.638	206.3	0.3197
250	17.91	0.705	251.9	0.3904
300	19.30	0.760	292.6	0.4536
350	20.60	0.811	333.3	0.5166
400	21.79	0.858	373.0	0.5782
500	23.95	0.943	450.6	0.6984
600	26.75	1.053	561.9	0.8709
700	28.55	1.124	640.2	0.9923
750	29.41	1.158	679.5	1.0532
800	30.23	1.190	717.5	1.1122
900	31.85	1.254	796.8	1.2351
1000	33.32	1.312	872.2	1.3519
1250	37.57	1.479	1108	1.7180
1500	40.69	1.602	1300	2.0157

Continued

Table 5 Dimensions of Insulated Conductors and Fixture Wires *Continued*

Type	Size (AWG or kcmil)	Approximate Diameter		Approximate Area	
		mm	in.	mm²	in.²
	Type: KF-1, KF-2, KF-1, KFF-2, XHH, XHHW, XHHW-2, ZW *Continued*				
XHHW, XHHW-2, XHH	1750	43.59	1.716	1492	2.3127
	2000	46.28	1.822	1682	2.6073
KF-2, KFF-2	18	1.600	0.063	2.000	0.0031
	16	1.905	0.075	2.839	0.0044
	14	2.286	0.090	4.129	0.0064
	12	2.769	0.109	6.000	0.0093
	10	3.378	0.133	8.968	0.0139
KF-1, KFF-1	18	1.448	0.057	1.677	0.0026
	16	1.753	0.069	2.387	0.0037
	14	2.134	0.084	3.548	0.0055
	12	2.616	0.103	5.355	0.0083
	10	3.226	0.127	8.194	0.0127

*Types RHH, RHW, and RHW-2 without outer covering.

Table 8 Conductor Properties

Size (AWG or kcmil)	Area		Stranding			Overall				DC Resistance at 75°C (167°F) Copper Uncoated		Copper Coated		Aluminum	
	mm²	Circular mils	Qty	Diameter mm	in.	Diameter mm	in.	Area mm²	in.²	ohm/km	ohm/kFT	ohm/km	ohm/kFT	ohm/km	ohm/kFT
18	0.823	1620	1	—	—	1.02	0.040	0.823	0.001	25.5	7.77	26.5	8.08	42.0	12.8
18	0.823	1620	7	0.39	0.015	1.16	0.046	1.06	0.002	26.1	7.95	27.7	8.45	42.8	13.1
16	1.31	2580	1	—	—	1.29	0.051	1.31	0.002	16.0	4.89	16.7	5.08	26.4	8.05
16	1.31	2580	7	0.49	0.019	1.46	0.058	1.68	0.003	16.4	4.99	17.3	5.29	26.9	8.21
14	2.08	4110	1	—	—	1.63	0.064	2.08	0.003	10.1	3.07	10.4	3.19	16.6	5.06
14	2.08	4110	7	0.62	0.024	1.85	0.073	2.68	0.004	10.3	3.14	10.7	3.26	16.9	5.17
12	3.31	6530	1	—	—	2.05	0.081	3.31	0.005	6.34	1.93	6.57	2.01	10.45	3.18
12	3.31	6530	7	0.78	0.030	2.32	0.092	4.25	0.006	6.50	1.98	6.73	2.05	10.69	3.25
10	5.261	10380	1	—	—	2.588	0.102	5.26	0.008	3.984	1.21	4.148	1.26	6.561	2.00
10	5.261	10380	7	0.98	0.308	2.95	0.116	6.76	0.011	4.070	1.24	4.226	1.29	6.679	2.04

Continued

Table 8 Conductor Properties *Continued*

Size (AWG or kcmil)	Area		Stranding			Overall				Direct-Current Resistance at 75°C (167°F)					
						Diameter		Area		Copper				Aluminum	
				Diameter						Uncoated		Coated			
	mm²	Circular mils	Qty	mm	in.	mm	in.	mm²	in.²	ohm/km	ohm/kFT	ohm/km	ohm/kFT	ohm/km	ohm/kFT
8	8.367	16510	1	—	—	3.264	0.128	8.37	0.013	2.506	0.764	2.579	0.786	4.125	1.26
8	8.367	16510	7	1.23	0.049	3.71	0.146	10.76	0.017	2.551	0.778	2.653	0.809	4.204	1.28
6	13.30	26240	7	1.56	0.061	4.67	0.184	17.09	0.027	1.608	0.491	1.671	0.510	2.652	0.808
4	21.15	41740	7	1.96	0.077	5.89	0.232	27.19	0.042	1.010	0.308	1.053	0.321	1.666	0.508
3	26.67	52620	7	2.20	0.087	6.60	0.260	34.28	0.053	0.802	0.245	0.833	0.254	1.320	0.403
2	33.62	66360	7	2.47	0.097	7.42	0.292	43.23	0.067	0.634	0.194	0.661	0.201	1.045	0.319
1	42.41	83690	19	1.69	0.066	8.43	0.332	55.80	0.087	0.505	0.154	0.524	0.160	0.829	0.253
1/0	53.49	105600	19	1.89	0.074	9.45	0.372	70.41	0.109	0.399	0.122	0.415	0.127	0.660	0.201
2/0	67.43	133100	19	2.13	0.084	10.62	0.418	88.74	0.137	0.3170	0.0967	0.329	0.101	0.523	0.159
3/0	85.01	167800	19	2.39	0.094	11.94	0.470	111.9	0.173	0.2512	0.0766	0.2610	0.0797	0.413	0.126
4/0	107.2	211600	19	2.68	0.106	13.41	0.528	141.1	0.219	0.1996	0.0608	0.2050	0.0626	0.328	0.100
250	—		37	2.09	0.082	14.61	0.575	168	0.260	0.1687	0.0515	0.1753	0.0535	0.2778	0.0847

300	—	37	2.29	0.090	16.00	0.630	201	0.312	0.1409	0.0429	0.1463	0.0446	0.2318	0.0707
350	—	37	2.47	0.097	17.30	0.681	235	0.364	0.1205	0.0367	0.1252	0.0382	0.1984	0.0605
400	—	37	2.64	0.104	18.49	0.728	268	0.416	0.1053	0.0321	0.1084	0.0331	0.1737	0.0529
500	—	37	2.95	0.116	20.65	0.813	336	0.519	0.0845	0.0258	0.0869	0.0265	0.1391	0.0424
600	—	61	2.52	0.099	22.68	0.893	404	0.626	0.0704	0.0214	0.0732	0.0223	0.1159	0.0353
700	—	61	2.72	0.107	24.49	0.964	471	0.730	0.0603	0.0184	0.0622	0.0189	0.0994	0.0303
750	—	61	2.82	0.111	25.35	0.998	505	0.782	0.0563	0.0171	0.0579	0.0176	0.0927	0.0282
800	—	61	2.91	0.114	26.16	1.030	538	0.834	0.0528	0.0161	0.0544	0.0166	0.0668	0.0265
900	—	61	3.09	0.122	27.79	1.094	606	0.940	0.0470	0.0143	0.0481	0.0147	0.0770	0.0235
1000	—	61	3.25	0.128	29.26	1.152	673	1.042	0.0423	0.0129	0.0434	0.0132	0.0695	0.0212
1250	—	91	2.98	0.117	32.74	1.289	842	1.305	0.0338	0.0103	0.0347	0.0106	0.0554	0.0169
1500	—	91	3.26	0.128	35.86	1.412	1011	1.566	0.02814	0.00858	0.02814	0.00883	0.0464	0.0141
1750	—	127	2.98	0.117	38.76	1.526	1180	1.829	0.02410	0.00735	0.02410	0.00756	0.0397	0.0121
2000	—	127	3.19	0.126	41.45	1.632	1349	2.092	0.02109	0.00643	0.02109	0.00662	0.0348	0.0106

Notes:

1. These resistance values are valid only for the parameters as given. Using conductors having coated strands, different stranding type, and, especially, other temperatures changes the resistance.

2. Formula for temperature change: $R_2 = R_1 [1 + \alpha(T_2 - 75)]$ where $\alpha_{Cu} = 0.00323$, $\alpha_M = 0.00330$ at 75°C.

3. Conductors with compact and compressed stranding have about 9 percent and 3 percent, respectively, smaller bare conductor diameters than those shown. See Table 5A for actual compact cable dimensions.

4. The IACS conductivities used: copper = 100%, aluminum = 61%.

5. Class B stranding is listed as well as solid for some sizes. Its overall diameter and area is that of its circumscribing circle.

ANNEX C

Conduit and Tubing Fill Tables for Conductors and Fixture Wires of the Same Size

Table C1 Maximum Number of Conductors or Fixture Wires in Electrical Metallic Tubing (EMT) (Based on Table 1)

CONDUCTORS

Type	Conductor Size (AWG/kcmil)	Metric Designator (Trade Size)									
		16 (1/2)	21 (3/4)	27 (1)	35 (1-1/4)	41 (1-1/2)	53 (2)	63 (2-1/2)	78 (3)	91 (3-1/2)	103 (4)
RHH, RHW, RHW-2	14	4	7	11	20	27	46	80	120	157	201
	12	3	6	9	17	23	38	66	100	131	167
	10	2	5	8	13	18	30	53	81	105	135
	8	1	2	4	7	9	16	28	42	55	70
	6	1	1	3	5	8	13	22	34	44	56
	4	1	1	2	4	6	10	17	26	34	44
	3	1	1	1	4	5	9	15	23	30	38
	2	1	1	1	3	4	7	13	20	26	33
	1	0	1	1	1	3	5	9	13	17	22
	1/0	0	1	1	2	2	4	7	11	15	19
	2/0	0	0	1	1	2	4	6	10	13	17
	3/0	0	0	1	1	1	3	5	8	11	14
	4/0	0	0	1	1	1	3	5	7	9	12

	C1	C2	C3	C4	C5	C6	C7	C8	C9	C10
250	9	7	5	3	1	1	1	0	0	0
300	8	6	5	3	1	1	1	0	0	0
350	7	6	4	3	1	1	1	0	0	0
400	7	5	4	2	1	1	1	0	0	0
500	6	4	3	2	1	1	0	0	0	0
600	5	4	3	1	1	1	0	0	0	0
700	4	3	2	1	1	0	0	0	0	0
750	4	3	2	1	1	0	0	0	0	0
800	4	3	2	1	1	0	0	0	0	0
900	3	3	1	1	1	0	0	0	0	0
1000	3	2	1	1	1	0	0	0	0	0
1250	2	1	1	1	0	0	0	0	0	0
1500	1	1	1	1	0	0	0	0	0	0
1750	1	1	1	1	0	0	0	0	0	0
2000	1	1	1	1	0	0	0	0	0	0
14	424	332	254	168	96	58	43	25	15	8
12	326	255	195	129	74	45	33	19	11	6
10	243	190	145	96	55	33	24	14	8	5
8	135	105	81	53	30	18	13	8	5	2

TW, THHW
THW, THW-2

Continued

Table C1 Maximum Number of Conductors or Fixture Wires in Electrical Metallic Tubing (EMT) (Based on Table 1) *Continued*

CONDUCTORS

Type	Conductor Size (AWG/kcmil)	Metric Designator (Trade Size)									
		16 (1/2)	21 (3/4)	27 (1)	35 (1-1/4)	41 (1-1/2)	53 (2)	63 (2-1/2)	78 (3)	91 (3-1/2)	103 (4)
RHH*, RHW*, RHW-2*	14	6	10	16	28	39	64	112	169	221	282
	12	4	8	13	23	31	51	90	136	177	227
	10	3	6	10	18	24	40	70	106	138	177
	8	1	4	6	10	14	24	42	63	83	106
RHH*, RHW*, RHW-2*, TW, THW, THHW, THW-2	6	1	3	4	8	11	18	32	48	63	81
	4	1	1	3	6	8	13	24	36	47	60
	3	1	1	3	5	7	12	20	31	40	52
	2	1	1	2	4	6	10	17	26	34	44
	1	1	1	1	3	4	7	12	18	24	31
	1/0	0	1	1	2	3	6	10	16	20	26
	2/0	0	0	1	1	3	5	9	13	17	22
	3/0	0	1	1	1	2	4	7	11	15	19
	4/0	0	0	1	1	1	3	6	9	12	16

250	0	0	1	1	1	3	5	7	10	13
300	0	0	1	1	1	2	4	6	8	11
350	0	0	0	1	1	1	4	6	7	10
400	0	0	0	1	1	1	3	5	7	9
500	0	0	0	1	1	1	3	4	6	7
600	0	0	0	1	1	1	2	3	4	6
700	0	0	0	0	0	1	1	3	4	5
750	0	0	0	0	0	1	1	3	4	5
800	0	0	0	0	0	1	1	3	3	5
900	0	0	0	0	0	1	1	2	3	4
1000	0	0	0	0	0	1	1	2	3	4
1250	0	0	0	0	0	1	1	1	2	3
1500	0	0	0	0	0	0	1	1	1	2
1750	0	0	0	0	0	0	1	1	1	2
2000	0	0	0	0	0	0	1	1	1	1
THHN, THWN, THWN-2 — 14	12	22	35	61	84	138	241	364	476	608
12	9	16	26	45	61	101	176	266	347	443
10	5	10	16	28	38	63	111	167	219	279

Continued

Table C1 Maximum Number of Conductors or Fixture Wires in Electrical Metallic Tubing (EMT) (Based on Table 1) *Continued*

CONDUCTORS

Type	Conductor Size (AWG/kcmil)	Metric Designator (Trade Size)									
		16 (1/2)	21 (3/4)	27 (1)	35 (1-1/4)	41 (1-1/2)	53 (2)	63 (2-1/2)	78 (3)	91 (3-1/2)	103 (4)
THHN, THWN, THWN-2	8	3	6	9	16	22	36	64	96	126	161
	6	2	4	7	12	16	26	46	69	91	116
	4	1	2	4	7	10	16	28	43	56	71
	3	1	1	3	6	8	13	24	36	47	60
	2	1	1	3	5	7	11	20	30	40	51
	1	1	1	1	4	5	8	15	22	29	37
	1/0	1	1	1	3	4	7	12	19	25	32
	2/0	0	1	1	2	3	6	10	16	20	26
	3/0	0	1	1	1	3	5	8	13	17	22
	4/0	0	1	1	1	2	4	7	11	14	18
	250	0	0	1	1	1	3	6	9	11	15
	300	0	0	1	1	1	3	5	7	10	13
	350	0	0	1	1	1	2	4	6	9	11

Type	Size										
	400	0	0	0	1	1	1	4	6	8	10
	500	0	0	0	1	1	1	3	5	6	8
	600	0	0	0	1	1	1	2	4	5	7
	700	0	0	0	1	1	1	2	3	4	6
	750	0	0	0	0	1	1	1	3	4	5
	800	0	0	0	0	1	1	1	3	4	5
	900	0	0	0	0	1	1	1	2	3	4
	1000	0	0	0	0	1	1	1	2	3	4
FEP, FEPB, PFA, PFAH, TFE	14	12	21	34	60	81	134	234	354	462	590
	12	9	15	25	43	59	98	171	258	337	430
	10	6	11	18	31	42	70	122	185	241	309
	8	3	6	10	18	24	40	70	106	138	177
	6	2	4	7	12	17	28	50	75	98	126
	4	1	3	5	9	12	20	35	53	69	88
	3	1	2	4	7	10	16	29	44	57	73
	2	1	1	3	6	8	13	24	36	47	60
PFA, PFAH, TFE	1	1	1	2	4	6	9	16	25	33	42

Continued

Table C1 Maximum Number of Conductors or Fixture Wires in Electrical Metallic Tubing (EMT) (Based on Table 1) *Continued*

CONDUCTORS

Type	Conductor Size (AWG/kcmil)	Metric Designator (Trade Size)									
		16 (1/2)	21 (3/4)	27 (1)	35 (1-1/4)	41 (1-1/2)	53 (2)	63 (2-1/2)	78 (3)	91 (3-1/2)	103 (4)
PFAH, TFE, PFA, PFAH, TFE, Z	1/0	1	1	1	3	5	8	14	21	27	35
	2/0	0	1	1	3	4	6	11	17	22	29
	3/0	0	1	1	2	3	5	9	14	18	24
	4/0	0	1	1	1	2	4	8	11	15	19
Z	14	14	25	41	72	98	161	282	426	556	711
	12	10	18	29	51	69	114	200	302	394	504
	10	6	11	18	31	42	70	122	185	241	309
	8	4	7	11	20	27	44	77	117	153	195
	6	3	5	8	14	19	31	54	82	107	137
	4	1	3	5	9	13	21	37	56	74	94
	3	1	2	4	7	9	15	27	41	54	69
	2	1	1	3	6	8	13	22	34	45	57
	1	1	1	2	4	6	10	18	28	36	46

XHH, XHHW, XHHW-2, ZW										
14	424	332	254	168	96	58	43	25	15	8
12	326	255	195	129	74	45	33	19	11	6
10	243	190	145	96	55	33	24	14	8	5
8	135	105	81	53	30	18	13	8	5	2
6	100	78	60	39	22	14	10	6	3	1
XHH, XHHW, XHHW-2										
4	72	56	43	28	16	10	7	4	2	1
3	61	48	36	24	14	8	6	3	1	1
2	51	40	31	20	11	7	5	3	1	1
1	38	30	23	15	8	5	4	1	1	1
1/0	32	25	19	13	7	4	3	1	1	1
2/0	27	21	16	10	6	3	2	1	1	0
3/0	22	17	13	9	5	3	1	1	1	0
4/0	18	14	11	7	4	2	1	1	1	0
250	15	12	9	6	3	1	1	1	0	0
300	13	10	8	5	3	1	1	1	0	0
350	11	9	7	4	2	1	1	1	0	0
400	10	8	6	4	1	1	1	0	0	0
500	8	6	5	3	1	1	1	0	0	0

Continued

Table C1 Maximum Number of Conductors or Fixture Wires in Electrical Metallic Tubing (EMT) (Based on Table 1) Continued

CONDUCTORS

Type	Conductor Size (AWG/kcmil)	Metric Designator (Trade Size)									
		16 (1/2)	21 (3/4)	27 (1)	35 (1-1/4)	41 (1-1/2)	53 (2)	63 (2-1/2)	78 (3)	91 (3-1/2)	103 (4)
XHH, XHHW, XHHW-2	600	0	0	0	1	1	1	2	4	5	6
	700	0	0	0	0	1	1	2	3	4	6
	750	0	0	0	0	1	1	1	3	4	5
	800	0	0	0	0	1	1	1	3	4	5
	900	0	0	0	0	1	1	1	3	3	4
	1000	0	0	0	0	0	1	1	2	3	4
	1250	0	0	0	0	0	1	1	1	2	3
	1500	0	0	0	0	0	1	1	1	1	3
	1750	0	0	0	0	0	0	1	1	1	2
	2000	0	0	0	0	0	0	1	1	1	1

Table C2 Maximum Number of Conductors or Fixture Wires in Electrical Nonmetallic Tubing (ENT) (Based on Table 1)

CONDUCTORS

Type	Conductor Size (AWG/kcmil)	Metric Designator (Trade Size)					
		16 (1/2)	21 (3/4)	27 (1)	35 (1-1/4)	41 (1-1/2)	53 (2)
RHH, RHW, RHW-2	14	3	6	10	19	26	43
	12	2	5	9	16	22	36
RHH, RHW, RHW-2	10	1	4	7	13	17	29
	8	1	1	3	6	9	15
	6	1	1	3	5	7	12
	4	1	1	2	4	6	9
	3	1	1	1	3	5	8
	2	0	1	1	3	4	7
	1	0	1	1	1	3	5
	1/0	0	0	1	1	2	4
	2/0	0	0	1	1	1	3

Continued

Table C2 Maximum Number of Conductors or Fixture Wires in Electrical Nonmetallic Tubing (ENT) (Based on Table 1) *Continued*

CONDUCTORS

Type	Conductor Size (AWG/kcmil)	Metric Designator (Trade Size)					
		16 (1/2)	21 (3/4)	27 (1)	35 (1-1/4)	41 (1-1/2)	53 (2)
RHH, RHW, RHW-2	3/0	0	0	1	1	1	3
	4/0	0	0	1	1	1	2
	250	0	0	0	1	1	1
	300	0	0	0	1	1	1
	350	0	0	0	1	1	1
	400	0	0	0	1	1	1
	500	0	0	0	0	1	1
	600	0	0	0	0	1	1
	700	0	0	0	0	0	1
	750	0	0	0	0	0	1
	800	0	0	0	0	0	1
	900	0	0	0	0	0	1

Type	Size						
	1000	0	0	0	0	0	1
	1250	0	0	0	0	0	0
	1500	0	0	0	0	0	0
	1750	0	0	0	0	0	0
	2000	0	0	0	0	0	0
TW, THHW, THW, THW-2	14	7	13	22	40	55	92
	12	5	10	17	31	42	71
	10	4	7	13	23	32	52
	8	1	4	7	13	17	29
RHH*, RHW*, RHW-2*	14	4	8	15	27	37	61
RHH*, RHW*, RHW-2*	12	3	7	12	21	29	49
	10	3	5	9	17	23	38
RHH*, RHW*, RHW-2*	8	1	3	5	10	14	23
RHH*, RHW*, RHW-2*, TW, THW, THHW, THW-2	6	1	2	4	7	10	17
	4	1	1	3	5	8	13
	3	1	1	2	5	7	11

Continued

Table C2 Maximum Number of Conductors or Fixture Wires in Electrical Nonmetallic Tubing (ENT) (Based on Table 1) *Continued*

Type	Conductor Size (AWG/kcmil)	CONDUCTORS Metric Designator (Trade Size)						
		16 (1/2)	21 (3/4)	27 (1)	35 (1-1/4)	41 (1-1/2)	53 (2)
RHH*, RHW*, RHW-2*, TW, THW, THHW, THW-2	2	1	1	2	4	6	9
	1	0	1	1	3	4	6
	1/0	0	1	1	2	3	5
	2/0	0	0	1	1	3	5
	3/0	0	0	1	1	2	4
	4/0	0	0	1	1	1	3
	250	0	0	1	1	1	2
	300	0	0	0	1	1	2
	350	0	0	0	1	1	1
	400	0	0	0	1	1	1
	500	0	0	0	0	1	1
	600	0	0	0	0	1	1
	700	0	0	0	0	1	1

750	0	0	0	0	0	1
800	0	0	0	0	1	1
900	0	0	0	0	0	1
1000	0	0	0	0	0	1
1250	0	0	0	0	0	1
1500	0	0	0	0	0	0
1750	0	0	0	0	0	0
2000	0	0	0	0	0	0
14	10	18	32	58	80	132
12	7	13	23	42	58	96
10	4	8	15	26	36	60
8	2	5	8	15	21	35
6	1	3	6	11	15	25
4	1	1	4	7	9	15
3	1	1	3	5	8	13
2	1	1	2	5	6	11
1	1	1	1	3	5	8
1/0	0	1	1	3	4	7
2/0	0	1	1	2	3	5

THHN, THWN, THWN-2

Continued

Table C2 Maximum Number of Conductors or Fixture Wires in Electrical Nonmetallic Tubing (ENT) (Based on Table 1) *Continued*

		CONDUCTORS							
					Metric Designator (Trade Size)				
Type	Conductor Size (AWG/kcmil)	16 (1/2)	21 (3/4)	27 (1)	35 (1-1/4)	41 (1-1/2)	53 (2)
THHN, THWN, THWN-2	3/0	0	1	1	1	3	4
	4/0	0	0	1	1	2	4
	250	0	0	1	1	1	3
	300	0	0	1	1	1	2
	350	0	0	0	1	1	2
	400	0	0	0	1	1	1
	500	0	0	0	1	1	1
	600	0	0	0	1	1	1
	700	0	0	0	0	1	1
	750	0	0	0	0	1	1
	800	0	0	0	0	1	1
	900	0	0	0	0	0	1
	1000	0	0	0	0	0	1

Type	Size						
FEP, FEPB, PFA, PFAH, TFE	14	10	18	31	56	77	128
	12	7	13	23	41	56	93
	10	5	9	16	29	40	67
	8	3	5	9	17	23	38
	6	1	4	6	12	16	27
PFA, PFAH, TFE	4	1	2	4	8	11	19
	3	1	1	4	7	9	16
	2	1	1	3	5	8	13
PFA, PFAH, TFE	1	1	1	1	4	5	9
PFA, PFAH, TFE, Z	1/0	0		1	3	4	7
	2/0	0		1	2	4	6
	3/0	0		1	1	3	5
	4/0	0		1	1	2	4
Z	14	12	22	38	68	93	154
	12	8	15	27	48	66	109

Continued

Table C2 Maximum Number of Conductors or Fixture Wires in Electrical Nonmetallic Tubing (ENT) (Based on Table 1) *Continued*

CONDUCTORS

Type	Conductor Size (AWG/kcmil)	Metric Designator (Trade Size)							
		16 (1/2)	21 (3/4)	27 (1)	35 (1-1/4)	41 (1-1/2)	53 (2)		
Z	10	5	9	16	29	40	67		
	8	3	6	10	18	25	42		
	6	1	4	7	13	18	30		
	4	1	3	5	9	12	20		
	3	1	1	3	6	9	15		
	2	1	1	3	5	7	12		
	1	1	1	2	4	6	10		
XHH, XHHW, XHHW-2, ZW	14	7	13	22	40	55	92		
	12	5	10	17	31	42	71		
	10	4	7	13	23	32	52		
	8	1	4	7	13	17	29		
	6	1	3	5	9	13	21		

Type	Size						
XHH, XHHW, XHHW-2, ZW	4	1	1	4	7	9	15
	3	1	1	3	6	8	13
	2	1	1	2	5	6	11
XHH, XHHW, XHHW-2	1	1	1	1	3	5	8
	1/0	0	1	1	3	4	7
	2/0	0	1	1	2	3	6
	3/0	0	1	1	1	3	5
	4/0	0	0	1	1	2	4
	250	0	0	1	1	1	3
	300	0	0	1	1	1	3
	350	0	0	1	1	1	2
	400	0	0	0	1	1	1
	500	0	0	0	1	1	1
	600	0	0	0	0	1	1
	700	0	0	0	0	1	1
	750	0	0	0	0	1	1
	800	0	0	0	0	1	1
	900	0	0	0	0	1	1

Continued

Table C2 Maximum Number of Conductors or Fixture Wires in Electrical Nonmetallic Tubing (ENT) (Based on Table 1) *Continued*

		CONDUCTORS						
		Metric Designator (Trade Size)						
Type	Conductor Size (AWG/kcmil)	16 (1/2)	21 (3/4)	27 (1)	35 (1-1/4)	41 (1-1/2)	53 (2)	
XHH, XHHW, XHHW-2	1000	0	0	0	0	0	1	
	1250	0	0	0	0	0	1	
	1500	0	0	0	0	0	1	
	1750	0	0	0	0	0	0	
	2000	0	0	0	0	0	0	

Table C3 Maximum Number of Conductors or Fixture Wires in Flexible Metal Conduit (FMC) (Based on Table 1)

| | | CONDUCTORS | | | | | | | | | |
| | | Metric Designator (Trade Size) | | | | | | | | | |
Type	Conductor Size (AWG/kcmil)	16 (1/2)	21 (3/4)	27 (1)	35 (1-1/4)	41 (1-1/2)	53 (2)	63 (2-1/2)	78 (3)	91 (3-1/2)	103 (4)
RHH, RHW, RHW-2	14	4	7	11	17	25	44	67	96	131	171
	12	3	6	9	14	21	37	55	80	109	142
RHH, RHW, RHW-2	10	3	5	7	11	17	30	45	64	88	115
	8	1	2	4	6	9	15	23	34	46	60
	6	1	1	3	5	7	12	19	27	37	48
	4	1	1	2	4	5	10	14	21	29	37
	3	1	1	1	3	5	8	13	18	25	33
	2	1	1	1	3	4	7	11	16	22	28
	1	0	1	1	1	2	5	7	10	14	19
	1/0	0	1	1	1	2	4	6	9	12	16
	2/0	0	1	1	1	1	3	5	8	11	14

Continued

Table C3 Maximum Number of Conductors or Fixture Wires in Flexible Metal Conduit (FMC) (Based on Table 1) *Continued*

CONDUCTORS

Type	Conductor Size (AWG/kcmil)	Metric Designator (Trade Size)									
		16 (1/2)	21 (3/4)	27 (1)	35 (1-1/4)	41 (1-1/2)	53 (2)	63 (2-1/2)	78 (3)	91 (3-1/2)	103 (4)
RHH, RHW, RHW-2	3/0	0	0	1	1	1	3	5	7	9	12
	4/0	0	0	1	1	1	2	4	6	8	10
	250	0	0	0	1	1	1	3	4	6	8
	300	0	0	0	1	1	1	2	4	5	7
	350	0	0	0	0	1	1	2	3	5	6
	400	0	0	0	0	1	1	1	3	4	6
	500	0	0	0	0	1	1	1	3	4	5
	600	0	0	0	0	0	1	1	2	3	4
	700	0	0	0	0	0	1	1	1	3	3
	750	0	0	0	0	0	1	1	1	2	3
	800	0	0	0	0	0	1	1	1	2	3
	900	0	0	0	0	0	1	1	1	2	3

Type	Size										
	1000	0	0	0	0	0	1	1	1	1	3
	1250	0	0	0	0	0	0	1	1	1	1
	1500	0	0	0	0	0	0	1	1	1	1
	1750	0	0	0	0	0	0	1	1	1	1
	2000	0	0	0	0	0	0	0	1	1	1
TW, THHW, THW, THW-2	14	9	15	23	36	53	94	141	203	277	361
	12	7	11	18	28	41	72	108	156	212	277
	10	5	8	13	21	30	54	81	116	158	207
	8	3	5	7	11	17	30	45	64	88	115
RHH*, RHW*, RHW-2*	14	6	10	15	24	35	62	94	135	184	240
RHH*, RHW*, RHW-2*	12	5	8	12	19	28	50	75	108	148	193
	10	4	6	10	15	22	39	59	85	115	151
RHH*, RHW*, RHW-2*	8	1	4	6	9	13	23	35	51	69	90

Continued

Table C3 Maximum Number of Conductors or Fixture Wires in Flexible Metal Conduit (FMC) (Based on Table 1) *Continued*

CONDUCTORS

Type	Conductor Size (AWG/kcmil)	Metric Designator (Trade Size)									
		16 (1/2)	21 (3/4)	27 (1)	35 (1-1/4)	41 (1-1/2)	53 (2)	63 (2-1/2)	78 (3)	91 (3-1/2)	103 (4)
RHH*, RHW*, RHW-2*, TW, THW, THW-2	6	1	3	4	7	10	18	27	39	53	69
	4	1	1	3	5	7	13	20	29	39	51
	3	1	1	3	4	6	11	17	25	34	44
	2	1	1	2	4	5	10	14	21	29	37
	1	1	1	1	2	4	7	10	15	20	26
	1/0	0	1	1	3	3	6	9	12	17	22
	2/0	0	1	1	1	3	5	7	10	14	19
	3/0	0	1	1	1	2	4	6	9	12	16
	4/0	0	0	1	1	1	3	5	7	10	13
	250	0	0	0	1	1	3	4	6	8	11
	300	0	0	1	1	1	2	3	5	7	9
	350	0	0	0	1	1	1	3	4	6	8

Type / Size (THHN, THWN, THWN-2)										
400	7	6	4	3	1	1	1	0	0	0
500	6	5	3	2	1	1	1	0	0	0
600	5	4	3	1	1	1	0	0	0	0
700	4	3	2	1	1	1	0	0	0	0
750	4	3	2	1	1	1	0	0	0	0
800	4	3	1	1	1	1	0	0	0	0
900	3	3	1	1	1	0	0	0	0	0
1000	3	2	1	1	1	0	0	0	0	0
1250	2	1	1	1	1	0	0	0	0	0
1500	1	1	1	1	0	0	0	0	0	0
1750	1	1	1	1	0	0	0	0	0	0
2000	1	1	1	1	0	0	0	0	0	0
14	518	396	291	202	134	76	52	33	22	13
12	378	289	212	147	98	56	38	24	16	9
10	238	182	134	93	62	35	24	15	10	6
8	137	105	77	53	35	20	14	9	6	3
6	99	76	55	38	25	14	10	6	4	2

Continued

Table C3 Maximum Number of Conductors or Fixture Wires in Flexible Metal Conduit (FMC) (Based on Table 1) *Continued*

CONDUCTORS

Type	Conductor Size (AWG/kcmil)	Metric Designator (Trade Size)										
		16 (1/2)	21 (3/4)	27 (1)	35 (1-1/4)	41 (1-1/2)	53 (2)	63 (2-1/2)	78 (3)	91 (3-1/2)	103 (4)
THHN, THWN, THWN-2	4	1	2	4	6	9	16	24	34	46	61
	3	1	1	3	5	7	13	20	29	39	51
	2	1	1	3	4	6	11	17	24	33	43
	1	1	1	1	3	4	8	12	18	24	32
	1/0	1	1	1	2	4	7	10	15	20	27
	2/0	0	1	1	1	3	6	9	12	17	22
	3/0	0	1	1	1	2	5	7	10	14	18
	4/0	0	1	1	1	1	4	6	8	12	15
	250	0	0	1	1	1	3	5	7	9	12
	300	0	0	1	1	1	3	4	6	8	11
	350	0	0	1	1	1	2	3	5	7	9
	400	0	0	0	1	1	1	3	5	6	8
	500	0	0	0	1	1	2	2	4	5	7

Type	Size										
	600	0	0	0	0	1	1	1	3	4	5
	700	0	0	0	0	1	1	1	3	4	5
	750	0	0	0	0	1	1	1	2	3	4
	800	0	0	0	0	1	1	1	2	3	4
	900	0	0	0	0	0	1	1	1	3	4
	1000	0	0	0	0	0	1	1	1	3	3
FEP, FEPB, PFA, PFAH, TFE	14	12	21	32	51	74	130	196	282	385	502
	12	9	15	24	37	54	95	143	206	281	367
	10	6	11	17	26	39	68	103	148	201	263
	8	4	6	10	15	22	39	59	85	115	151
	6	2	4	7	11	16	28	42	60	82	107
	4	1	3	5	7	11	19	29	42	57	75
	3	1	2	4	6	9	16	24	35	48	62
	2	1	1	3	5	7	13	20	29	39	51
PFA, PFAH, TFE	1	1	1	2	3	5	9	14	20	27	36
PFA, PFAH, TFE, Z	1/0	1	1	1	3	4	8	11	17	23	30
	2/0	1	1	1	2	3	6	9	14	19	24

Continued

Table C3 Maximum Number of Conductors or Fixture Wires in Flexible Metal Conduit (FMC) (Based on Table 1) *Continued*

CONDUCTORS

Type	Conductor Size (AWG/kcmil)	Metric Designator (Trade Size)										
		16 (1/2)	21 (3/4)	27 (1)	35 (1-1/4)	41 (1-1/2)	53 (2)	63 (2-1/2)	78 (3)	91 (3-1/2)	103 (4)
PFA, PFAH, TFE, Z	3/0	0	1	1	1	3	5	8	11	15	20
	4/0	0	1	1	1	2	4	6	9	13	16
Z	14	15	25	39	61	89	157	236	340	463	605
	12	11	18	28	43	63	111	168	241	329	429
	10	6	11	17	26	39	68	103	148	201	263
	8	4	7	11	17	24	43	65	93	127	166
	6	3	5	7	12	17	30	45	65	89	117
	4	1	3	5	8	12	21	31	45	61	80
	3	1	2	4	6	8	15	23	33	45	58
	2	1	1	3	5	7	12	19	27	37	49
	1	1	1	2	4	6	10	15	22	30	39

XHH, XHHW, XHHW-2, ZW	14	9	15	23	36	53	94	141	203	277	361
	12	7	11	18	28	41	72	108	156	212	277
	10	5	8	13	21	30	54	81	116	158	207
	8	3	5	7	11	17	30	45	64	88	115
	6	1	3	5	8	12	22	33	48	65	85
	4	1	2	4	6	9	16	24	34	47	61
	3	1	1	3	5	7	13	20	29	40	52
	2	1	1	3	4	6	11	17	24	33	44
XHH, XHHW, XHHW-2	1/0	1	1	1	3	5	8	13	18	25	32
	2/0	1	1	1	2	4	7	10	15	21	27
	3/0	0	1	1	2	3	6	9	13	17	23
	4/0	0	1	1	1	2	5	7	10	14	19
	250	0	0	1	1	1	3	5	7	10	13
	300	0	0	1	1	1	3	4	6	8	11
	350	0	0	1	1	1	2	4	5	7	9
	400	0	0	0	1	1	1	3	5	6	8
	500	0	0	0	1	1	1	3	4	5	7

Continued

Table C3 Maximum Number of Conductors or Fixture Wires in Flexible Metal Conduit (FMC) (Based on Table 1) *Continued*

CONDUCTORS

Type	Conductor Size (AWG/kcmil)	Metric Designator (Trade Size)									
		16 (1/2)	21 (3/4)	27 (1)	35 (1-1/4)	41 (1-1/2)	53 (2)	63 (2-1/2)	78 (3)	91 (3-1/2)	103 (4)
XHH, XHHW, XHHW-2	600	0	0	0	0	1	1	1	3	4	5
	700	0	0	0	0	1	1	1	3	4	5
	750	0	0	0	0	1	1	1	2	3	4
	800	0	0	0	0	1	1	1	2	3	4
	900	0	0	0	0	0	1	1	1	3	4
	1000	0	0	0	0	0	1	1	1	3	3
	1250	0	0	0	0	0	1	1	1	1	3
	1500	0	0	0	0	0	1	1	1	1	2
	1750	0	0	0	0	0	0	1	1	1	1
	2000	0	0	0	0	0	0	1	1	1	1

*Types RHH, RHW, and RHW-2 without outer covering.

Conduit and Tubing Fill Tables

Table C4 Maximum Number of Conductors or Fixture Wires in Intermediate Metal Conduit (IMC) (Based on Table 1)

Type	Conductor Size (AWG/kcmil)	CONDUCTORS — Metric Designator (Trade Size)									
		16 (1/2)	21 (3/4)	27 (1)	35 (1-1/4)	41 (1-1/2)	53 (2)	63 (2-1/2)	78 (3)	91 (3-1/2)	103 (4)
RHH, RHW, RHW-2	14	4	8	13	22	30	49	70	108	144	186
	12	4	6	11	18	25	41	58	89	120	154
RHH, RHW, RHW-2	10	3	5	8	15	20	33	47	72	97	124
	8	1	3	4	8	10	17	24	38	50	65
	6	1	1	3	6	8	14	19	30	40	52
	4	1	1	3	5	6	11	15	23	31	41
	3	1	1	2	4	6	9	13	21	28	36
	2	1	1	2	3	5	8	11	18	24	31
	1	0	1	1	2	3	5	7	12	16	20
	1/0	0	1	1	1	3	4	6	10	14	18
	2/0	0	1	1	1	2	4	6	9	12	15

Continued

Table C4 Maximum Number of Conductors or Fixture Wires in Intermediate Metal Conduit (IMC) (Based on Table 1) *Continued*

		CONDUCTORS									
	Conductor Size (AWG/kcmil)	Metric Designator (Trade Size)									
Type		16 (1/2)	21 (3/4)	27 (1)	35 (1-1/4)	41 (1-1/2)	53 (2)	63 (2-1/2)	78 (3)	91 (3-1/2)	103 (4)
RHH, RHW, RHW-2	3/0	0	0	1	1	1	3	5	7	10	13
	4/0	0	0	1	1	1	3	4	6	9	11
	250	0	0	1	1	1	1	3	5	6	8
	300	0	0	0	1	1	1	3	4	6	7
	350	0	0	0	1	1	1	2	4	5	7
	400	0	0	0	1	1	1	2	3	5	6
	500	0	0	0	1	1	1	1	3	4	5
	600	0	0	0	0	1	1	1	2	3	4
	700	0	0	0	0	1	1	1	2	3	4
	750	0	0	0	0	1	1	1	1	3	4
	800	0	0	0	0	0	1	1	1	3	3
	900	0	0	0	0	0	1	1	1	2	3

Type	Size														
	1000	0	0	0	0	0	0	0	0	0	1	1	1	2	3
	1250	0	0	0	0	0	0	0	0	0	1	1	1	1	2
	1500	0	0	0	0	0	0	0	0	0	0	1	1	1	1
	1750	0	0	0	0	0	0	0	0	0	0	1	1	1	1
	2000	0	0	0	0	0	0	0	0	0	0	1	1	1	1
TW, THHW, THW, THW-2	14	0	0	0	0	10	17	27	47	64	104	147	228	304	392
	12	0	0	0	0	7	13	21	36	49	80	113	175	234	301
	10	0	0	0	0	5	9	15	27	36	59	84	130	174	224
	8	0	0	0	0	3	5	8	15	20	33	47	72	97	124
RHH*, RHW*, RHW-2*	14	0	0	0	0	6	11	18	31	42	69	98	151	202	261
RHH*, RHW*, RHW-2*	12	0	0	0	0	5	9	14	25	34	56	79	122	163	209
	10	0	0	0	0	4	7	11	19	26	43	61	95	127	163
RHH*, RHW*, RHW-2*	8	0	0	0	0	2	4	7	12	16	26	37	57	76	98
RHH*, RHW*, RHW-2*	6	0	0	0	0	1	3	5	9	12	20	28	43	58	75
	4	0	0	0	0	1	2	4	6	9	15	21	32	43	56

Continued

Table C4 Maximum Number of Conductors or Fixture Wires in Intermediate Metal Conduit (IMC) (Based on Table 1) *Continued*

	CONDUCTORS									
	Metric Designator (Trade Size)									
Conductor Size (AWG/kcmil)	16 (1/2)	21 (3/4)	27 (1)	35 (1-1/4)	41 (1-1/2)	53 (2)	63 (2-1/2)	78 (3)	91 (3-1/2)	103 (4)
Type TW, THW, THHW, THW-2										
3	1	1	3	6	8	13	18	28	37	48
2	1	1	3	5	6	11	15	23	31	41
1	1	1	1	3	4	7	11	16	22	28
1/0	1	1	1	3	4	6	9	14	19	24
2/0	0	1	1	2	3	5	8	12	16	20
3/0	0	1	1	1	3	4	6	10	13	17
4/0	0	1	1	1	2	4	5	8	11	14
250	0	0	1	1	1	3	4	7	9	12
300	0	0	1	1	2	2	4	6	8	10
350	0	0	1	1	2	2	3	5	7	9
400	0	0	0	1	1	1	3	4	6	8
500	0	0	0	1	1	1	2	4	5	7

Size										
600	0	0	0	1	1	1	1	3	4	5
700	0	0	0	0	1	1	1	3	4	5
750	0	0	0	0	1	1	1	2	3	4
800	0	0	0	0	1	1	1	2	3	4
900	0	0	0	0	1	1	1	2	3	4
1000	0	0	0	0	0	1	1	1	3	3
1250	0	0	0	0	0	1	1	1	1	3
1500	0	0	0	0	0	1	1	1	1	2
1750	0	0	0	0	0	0	1	1	1	1
2000	0	0	0	0	0	0	1	1	1	1

THHN, THWN, THWN-2

Size										
14	14	24	39	68	91	149	211	326	436	562
12	10	17	29	49	67	109	154	238	318	410
10	6	11	18	31	42	68	97	150	200	258
8	3	6	10	18	24	39	56	86	115	149
6	2	4	7	13	17	28	40	62	83	107
4	1	3	4	8	10	17	25	38	51	66
3	1	2	4	6	9	15	21	32	43	56

Continued

Table C4 Maximum Number of Conductors or Fixture Wires in Intermediate Metal Conduit (IMC) (Based on Table 1) *Continued*

CONDUCTORS

Type	Conductor Size (AWG/kcmil)	Metric Designator (Trade Size)									
		16 (1/2)	21 (3/4)	27 (1)	35 (1-1/4)	41 (1-1/2)	53 (2)	63 (2-1/2)	78 (3)	91 (3-1/2)	103 (4)
THHN, THWN, THWN-2	2	1	1	3	5	7	12	17	27	36	47
	1	1	1	2	4	5	9	13	20	27	35
	1/0	1	1	1	3	4	8	11	17	23	29
	2/0	1	1	1	3	4	6	9	14	19	24
	3/0	0	1	1	2	3	5	7	12	16	20
	4/0	0	0	1	1	2	4	6	9	13	17
	250	0	0	1	1	1	3	5	8	10	13
	300	0	0	1	1	1	3	4	7	9	12
	350	0	0	1	1	1	2	4	6	8	10
	400	0	0	1	1	1	2	3	5	7	9
	500	0	0	0	1	1	1	3	4	6	7

Type	Size										
	600	0	0	0	1	1	1	2	3	5	6
	700	0	0	0	1	1	1	1	3	4	5
	750	0	0	0	1	1	1	1	3	4	5
	800	0	0	0	0	1	1	1	3	4	5
	900	0	0	0	0	1	1	1	2	3	4
	1000	0	0	0	0	1	1	1	2	3	4
FEP, FEPB, PFA, PFAH, TFE	14	13	23	38	66	89	145	205	317	423	545
	12	10	17	28	48	65	106	150	231	309	398
	10	7	12	20	34	46	76	107	166	221	285
	8	4	7	11	19	26	43	61	95	127	163
	6	3	5	8	14	19	31	44	67	90	116
	4	1	3	5	10	13	21	30	47	63	81
	3	1	3	4	8	11	18	25	39	52	68
	2	1	2	4	6	9	15	21	32	43	56
PFA, PFAH, TFE	1	1	1	2	4	6	10	14	22	30	39
PFA, PFAH, TFE, Z	1/0	1	1	1	4	5	8	12	19	25	32
	2/0	1	1	1	3	4	7	10	15	21	27

Continued

Table C4 Maximum Number of Conductors or Fixture Wires in Intermediate Metal Conduit (IMC) (Based on Table 1) *Continued*

CONDUCTORS

Type	Conductor Size (AWG/kcmil)	16 (1/2)	21 (3/4)	27 (1)	35 (1-1/4)	41 (1-1/2)	53 (2)	63 (2-1/2)	78 (3)	91 (3-1/2)	103 (4)
						Metric Designator (Trade Size)					
PFA, PFAH, TFE, Z	3/0	0	1	1	2	3	6	8	13	17	22
	4/0	0	1	1	1	3	5	7	10	14	18
Z	14	16	28	46	79	107	175	247	381	510	657
	12	11	20	32	56	76	124	175	271	362	466
	10	7	12	20	34	46	76	107	166	221	285
	8	4	7	12	21	29	48	68	105	140	180
	6	3	5	9	15	20	33	47	73	98	127
	4	1	3	6	10	14	23	33	50	67	87
	3	1	2	4	7	10	17	24	37	49	63
	2	1	1	3	6	8	14	20	30	41	53
	1	1	1	3	5	7	11	16	25	33	43

XHH, XHHW, XHHW-2, ZW

	10	17	27	47	64	104	147	228	304	392
14	10	17	27	47	64	104	147	228	304	392
12	7	13	21	36	49	80	113	175	234	301
10	5	9	15	27	36	59	84	130	174	224
8	3	5	8	15	20	33	47	72	97	124
6	1	4	6	11	15	24	35	53	71	92
4	1	3	4	8	11	18	25	39	52	67
3	1	2	4	7	9	15	21	33	44	56
2	1	1	3	5	7	12	18	27	37	47

XHH, XHHW-2

	10	17	27	47	64	104	147	228	304	392
1	1	1	2	4	5	9	13	20	27	35
1/0	1	1	1	3	5	8	11	17	23	30
2/0	1	1	1	3	4	6	9	14	19	25
3/0	0	1	1	2	3	5	7	12	16	20
4/0	0	1	1	1	2	4	6	10	13	17
250	0	0	1	1	1	3	5	8	11	14
300	0	0	1	1	1	3	4	7	9	12
350	0	0	1	1	1	3	4	6	8	10
400	0	0	1	1	1	2	3	5	7	9
500	0	0	0	1	1	1	3	4	6	8

Continued

Table C4 Maximum Number of Conductors or Fixture Wires in Intermediate Metal Conduit (IMC) (Based on Table 1) *Continued*

CONDUCTORS

Type	Conductor Size (AWG/kcmil)	Metric Designator (Trade Size)									
		16 (1/2)	21 (3/4)	27 (1)	35 (1-1/4)	41 (1-1/2)	53 (2)	63 (2-1/2)	78 (3)	91 (3-1/2)	103 (4)
XHH, XHHW, XHHW-2	600	0	0	0	1	1	1	2	3	5	6
	700	0	0	0	1	1	1	1	3	4	5
	750	0	0	0	1	1	1	1	3	4	5
	800	0	0	0	0	1	1	1	3	4	5
	900	0	0	0	0	1	1	1	2	3	4
	1000	0	0	0	0	1	1	1	2	3	4
	1250	0	0	0	0	0	1	1	1	2	3
	1500	0	0	0	0	0	1	1	1	1	2
	1750	0	0	0	0	0	1	1	1	1	2
	2000	0	0	0	0	0	0	1	1	1	1

Table C5 Maximum Number of Conductors or Fixture Wires in Liquidtight Flexible Nonmetallic Conduit (Type LFNC-B*) (Based on Table 1)

	Conductor Size (AWG/kcmil)	CONDUCTORS							
		Metric Designator (Trade Size)							
Type		12 (3/8)	16 (1/2)	21 (3/4)	27 (1)	35 (1-1/4)	41 (1-1/2)	53 (2)
RHH, RHW, RHW-2	14	2	4	7	12	21	27	44
	12	1	3	6	10	17	22	36
RHH, RHW, RHW-2	10	1	3	5	8	14	18	29
	8	1	1	2	4	7	9	15
	6	1	1	1	3	6	7	12
	4	0	1	1	2	4	6	9
	3	0	1	1	1	4	5	8
	2	0	1	1	1	3	4	7
	1	0	0	1	1	1	3	5
	1/0	0	0	1	1	1	2	4
	2/0	0	0	1	1	1	1	3

Continued

Table C5 Maximum Number of Conductors or Fixture Wires in Liquidtight Flexible Nonmetallic Conduit (Type LFNC-B*) (Based on Table 1) *Continued*

CONDUCTORS

Type	Conductor Size (AWG/kcmil)	Metric Designator (Trade Size)						
		12 (3/8)	16 (1/2)	21 (3/4)	27 (1)	35 (1-1/4)	41 (1-1/2)	53 (2)
RHH, RHW, RHW-2	3/0	0	0	0	1	1	1	3
	4/0	0	0	0	1	1	1	2
	250	0	0	0	0	1	1	1
	300	0	0	0	0	1	1	1
	350	0	0	0	0	1	1	1
	400	0	0	0	0	1	1	1
	500	0	0	0	0	1	1	1
	600	0	0	0	0	0	1	1
	700	0	0	0	0	0	0	1
	750	0	0	0	0	0	0	1
	800	0	0	0	0	0	0	1
	900	0	0	0	0	0	0	1

Type	Size							
	1000	0	0	0	0	0	0	1
	1250	0	0	0	0	0	0	0
	1500	0	0	0	0	0	0	0
	1750	0	0	0	0	0	0	0
	2000	0	0	0	0	0	0	0
TW, THHW, THW, THW-2	14	5	9	15	25	44	57	93
	12	4	7	12	19	33	43	71
	10	3	5	9	14	25	32	53
	8	1	3	5	8	14	18	29
RHH†, RHW†, RHW-2†	14	3	6	10	16	29	38	62
RHH†, RHW†, RHW-2†	12	3	5	8	13	23	30	50
	10	1	3	6	10	18	23	39
RHH†, RHW†, RHW-2†	8	1	1	4	6	11	14	23

Continued

Table C5 Maximum Number of Conductors or Fixture Wires in Liquidtight Flexible Nonmetallic Conduit (Type LFNC-B*) (Based on Table 1) *Continued*

Type	Conductor Size (AWG/kcmil)	CONDUCTORS							
		Metric Designator (Trade Size)							
		12 (3/8)	16 (1/2)	21 (3/4)	27 (1)	35 (1-1/4)	41 (1-1/2)	53 (2)
RHH†, RHW†, RHW-2†, TW, THW	6	1	1	3	5	8	11	18
	4	1	1	1	3	6	8	13
	3	1	1	1	3	5	7	11
THHW, THW-2	2	0	1	1	2	4	6	9
	1	0	1	1	1	3	4	7
	1/0	0	0	1	1	2	3	6
	2/0	0	0	1	1	1	3	5
	3/0	0	0	1	1	1	2	4
	4/0	0	0	0	1	1	1	3
	250	0	0	0	1	1	1	3
	300	0	0	0	1	1	1	2
	350	0	0	0	0	1	1	1

400	0	0	0	0	1	1	1
500	0	0	0	0	1	1	1
600	0	0	0	0	1	1	1
700	0	0	0	0	0	1	1
750	0	0	0	0	0	1	1
800	0	0	0	0	0	1	1
900	0	0	0	0	0	1	1
1000	0	0	0	0	0	0	1
1250	0	0	0	0	0	0	1
1500	0	0	0	0	0	0	0
1750	0	0	0	0	0	0	0
2000	0	0	0	0	0	0	0
14	8	13	22	36	63	81	133
12	5	9	16	26	46	59	97
10	3	6	10	16	29	37	61
8	1	3	6	9	16	21	35
6	1	2	4	7	12	15	25

THHN, THWN, THWN-2

Continued

Table C5 Maximum Number of Conductors or Fixture Wires in Liquidtight Flexible Nonmetallic Conduit (Type LFNC-B*) (Based on Table 1). *Continued*

CONDUCTORS

Type	Conductor Size (AWG/kcmil)	Metric Designator (Trade Size)						
		12 (3/8)	16 (1/2)	21 (3/4)	27 (1)	35 (1-1/4)	41 (1-1/2)	53 (2)
THHN, THWN, THWN-2	4	1	1	2	4	7	9	15
	3	1	1	1	3	6	8	13
	2	1	1	1	3	5	7	11
	1	0	1	1	1	4	5	8
	1/0	0	0	1	1	3	4	7
	2/0	0	0	1	1	2	3	6
	3/0	0	0	1	1	1	3	5
	4/0	0	0	1	1	1	2	4
	250	0	0	0	1	1	1	3
	300	0	0	0	1	1	1	3
	350	0	0	0	1	1	1	2
	400	0	0	0	0	1	1	1
	500	0	0	0	0	1	1	1

Type	Size							
	600	1	1	1	0	0	0	0
	700	1	1	1	0	0	0	0
	750	1	1	0	0	0	0	0
	800	1	1	0	0	0	0	0
	900	1	1	0	0	0	0	0
	1000	1	0	0	0	0	0	0
FEP, FEPB, PFA, PFAH, TFE	14	129	79	61	35	21	12	7
	12	94	57	44	25	15	9	5
	10	68	41	32	18	11	6	4
	8	39	23	18	10	6	3	1
	6	27	17	13	7	4	2	1
	4	19	12	9	5	3	1	1
	3	16	10	7	4	2	1	1
	2	13	8	6	3	1	1	1
PFA, PFAH, TFE	1	9	5	4	2	1	1	0
PFA, PFAH	1/0	7	4	3	1	1	1	0

Continued

Table C5 Maximum Number of Conductors or Fixture Wires in Liquidtight Flexible Nonmetallic Conduit (Type LFNC-B*) (Based on Table 1) *Continued*

CONDUCTORS

Type	Conductor Size (AWG/kcmil)	Metric Designator (Trade Size)							
		12 (3/8)	16 (1/2)	21 (3/4)	27 (1)	35 (1-1/4)	41 (1-1/2)	53 (2)
TFE, Z	2/0	0	1	1	1	3	4	6
	3/0	0	0	1	1	2	3	5
	4/0	0	0	1	1	1	2	4
Z	14	9	15	26	42	73	95	156
	12	6	10	18	30	52	67	111
	10	4	6	11	18	32	41	68
	8	2	4	7	11	20	26	43
	6	1	3	5	8	14	18	30
	4	1	1	3	5	9	12	20
	3	1	1	2	4	7	9	15
	2	0	1	1	3	6	7	12
	1	0	1	1	2	5	6	10

Type	Size							
XHH, XHHW, XHHW-2, ZW	14	5	9	15	25	44	57	93
	12	4	7	12	19	33	43	71
	10	3	5	9	14	25	32	53
	8	1	3	5	8	14	18	29
	6	1	1	3	6	10	13	22
	4	1	1	2	4	7	9	16
	3	1	1	1	3	6	8	13
	2	1	1	1	3	5	7	11
	1	0	1	1	1	4	5	8
XHH, XHHW, XHHW-2	1/0	0	1	1	1	3	4	7
	2/0	0	0	1	1	2	3	6
	3/0	0	0	1	1	1	3	5
	4/0	0	0	1	1	1	2	4
	250	0	0	0	1	1	1	3
	300	0	0	0	1	1	1	3
	350	0	0	0	1	1	1	2

Continued

Table C5 Maximum Number of Conductors or Fixture Wires in Liquidtight Flexible Nonmetallic Conduit (Type LFNC-B*) (Based on Table 1) *Continued*

CONDUCTORS

Type	Conductor Size (AWG/kcmil)	Metric Designator (Trade Size)						
		12 (3/8)	16 (1/2)	21 (3/4)	27 (1)	35 (1-1/4)	41 (1-1/2)	53 (2)
XHH, XHHW, XHHW-2	400	0	0	0	0	1	1	1
	500	0	0	0	0	1	1	1
	600	0	0	0	0	1	1	1
	700	0	0	0	0	1	1	1
	750	0	0	0	0	0	1	1
	800	0	0	0	0	0	1	1
	900	0	0	0	0	0	1	1
	1000	0	0	0	0	0	0	1
	1250	0	0	0	0	0	0	1
	1500	0	0	0	0	0	0	1
	1750	0	0	0	0	0	0	0
	2000	0	0	0	0	0	0	0

Table C6 Maximum Number of Conductors or Fixture Wires in Liquidtight Flexible Nonmetallic Conduit (Type LFNC-A*) (Based on Table 1)

CONDUCTORS

Type	Conductor Size (AWG/kcmil)	Metric Designator (Trade Size)						
		12 (3/8)	16 (1/2)	21 (3/4)	27 (1)	35 (1-1/4)	41 (1-1/2)	53 (2)
RHH, RHW, RHW-2	14	2	4	7	11	20	27	45
	12	1	3	6	9	17	23	38
	10	1	3	5	8	13	18	30
	8	1	1	2	4	7	9	16
	6	1	1	1	3	5	7	13
	4	0	1	1	2	4	6	10
	3	0	1	1	1	4	5	8
	2	0	1	1	1	3	4	7
	1	0	0	1	1	1	3	5
	1/0	0	0	1	1	1	2	4
	2/0	0	0	1	1	1	1	4

Continued

Table C6 Maximum Number of Conductors or Fixture Wires in Liquidtight Flexible Nonmetallic Conduit (Type LFNC-A*) (Based on Table 1) *Continued*

CONDUCTORS

Type	Conductor Size (AWG/kcmil)	Metric Designator (Trade Size)								
		12 (3/8)	16 (1/2)	21 (3/4)	27 (1)	35 (1-1/4)	41 (1-1/2)	53 (2)
RHH, RHW, RHW-2	3/0	0	0	0	1	1	1	3
	4/0	0	0	0	1	1	1	3
	250	0	0	0	0	1	1	1
	300	0	0	0	0	1	1	1
	350	0	0	0	0	1	1	1
	400	0	0	0	0	1	1	1
	500	0	0	0	0	0	1	1
	600	0	0	0	0	0	1	1
	700	0	0	0	0	0	0	1
	750	0	0	0	0	0	0	1
	800	0	0	0	0	0	0	1
	900	0	0	0	0	0	0	1

Type	Size							
	1000	0	0	0	0	0	0	1
	1250	0	0	0	0	0	0	0
	1500	0	0	0	0	0	0	0
	1750	0	0	0	0	0	0	0
	2000	0	0	0	0	0	0	0
TW, THHW, THW, THW-2	14	5	9	15	24	43	58	96
	12	4	7	12	19	33	44	74
	10	3	5	9	14	24	33	55
	8	1	3	5	8	13	18	30
RHH†, RHW†, RHW-2†	14	3	6	10	16	28	38	64
	12	3	4	8	13	23	31	51
	10	1	3	6	10	18	24	40
	8	1	1	4	6	10	14	24
RHH†, RHW†, RHW-2†, TW, THW, THHW, THW-2	6	1	1	3	4	8	11	18
	4	1	1	1	3	6	8	13
	3	1	1	1	3	5	7	11
	2	0	1	1	2	4	6	10
	1	0	1	1	1	3	4	7

Continued

Table C6 Maximum Number of Conductors or Fixture Wires in Liquidtight Flexible Nonmetallic Conduit (Type LFNC-A*) (Based on Table 1) *Continued*

CONDUCTORS

Type	Conductor Size (AWG/kcmil)	Metric Designator (Trade Size)						
		12 (3/8)	16 (1/2)	21 (3/4)	27 (1)	35 (1-1/4)	41 (1-1/2)	53 (2)
RHH†, RHW†, RHW-2†, TW, THW, THHW, THW-2	1/0	0	0	0	1	2	3	6
	2/0	0	0	1	1	1	3	5
	3/0	0	0	1	1	1	2	4
	4/0	0	0	0	1	1	1	3
	250	0	0	0	1	1	1	3
	300	0	0	0	1	1	1	2
	350	0	0	0	0	1	1	1
	400	0	0	0	0	1	1	1
	500	0	0	0	0	1	1	1
	600	0	0	0	0	0	1	1
	700	0	0	0	0	0	1	1
	750	0	0	0	0	0	0	1
	800	0	0	0	0	0	0	1
	900	0	0	0	0	0	0	1

THHN, THWN, THWN-2

Size							
1000	0	0	0	0	0	0	1
1250	0	0	0	0	0	0	1
1500	0	0	0	0	0	0	0
1750	0	0	0	0	0	0	0
2000	0	0	0	0	0	0	0
14	8	13	22	35	62	83	137
12	5	9	16	25	45	60	100
10	3	6	10	16	28	38	63
8	1	3	6	9	16	22	36
6	1	2	4	6	12	16	26
4	1	1	2	4	7	9	16
3	1	1	1	3	6	8	13
2	1	1	1	3	5	7	11
1	0	1	1	1	4	5	8
1/0	0	1	1	1	3	4	7
2/0	0	0	1	1	2	3	6
3/0	0	0	1	1	1	3	5
4/0	0	0	1	1	1	2	4

Continued

Table C6 Maximum Number of Conductors or Fixture Wires in Liquidtight Flexible Nonmetallic Conduit (Type LFNC-A*) (Based on Table 1) *Continued*

Type	Conductor Size (AWG/kcmil)	CONDUCTORS Metric Designator (Trade Size)						
		12 (3/8)	16 (1/2)	21 (3/4)	27 (1)	35 (1-1/4)	41 (1-1/2)	53 (2)
THHN, THWN, THWN-2	250	0	0	0	1	1	1	3
	300	0	0	0	1	1	1	3
	350	0	0	0	1	1	1	2
	400	0	0	0	0	1	1	1
	500	0	0	0	0	1	1	1
	600	0	0	0	0	1	1	1
	700	0	0	0	0	0	1	1
	750	0	0	0	0	0	1	1
	800	0	0	0	0	0	1	1
	900	0	0	0	0	0	0	1
	1000	0	0	0	0	0	0	1
FEP, FEPB, PFA, PFAH, TFE	14	7	12	21	34	60	80	133
	12	5	9	15	25	44	59	97

Type	Size							
	10	70	42	31	18	11	6	4
	8	40	24	18	10	6	3	1
	6	28	17	13	7	4	2	1
	4	20	12	9	5	3	1	1
	3	16	10	7	4	2	1	1
	2	13	8	6	3	1	1	1
PFA, PFAH, TFE	1	9	5	4	2	1	1	0
PFA, PFAH, TFE, Z	1/0	8	5	3	1	1	1	0
	2/0	6	4	3	1	1	1	0
	3/0	5	3	2	1	1	0	0
	4/0	4	2	1	1	1	0	0
Z	14	161	97	72	41	25	15	9
	12	114	69	51	29	18	10	6
	10	70	42	31	18	11	6	4
	8	44	26	20	11	7	4	2
	6	31	18	14	8	5	3	1

Continued

Table C6 Maximum Number of Conductors or Fixture Wires in Liquidtight Flexible Nonmetallic Conduit (Type LFNC-A*) (Based on Table 1) *Continued*

CONDUCTORS

Type	Conductor Size (AWG/kcmil)	Metric Designater (Trade Size)								
		12 (3/8)	16 (1/2)	21 (3/4)	27 (1)	35 (1-1/4)	41 (1-1/2)	53 (2)		
Z	4	1	1	3	5	9	13	21		
	3	1	1	2	4	7	9	15		
	2	1	1	1	3	6	8	13		
	1	1	1	1	2	4	6	10		
XHH, XHHW, XHHW-2, ZW	14	5	9	15	24	43	58	96		
	12	4	7	12	19	33	44	74		
	10	3	5	9	14	24	33	55		
	8	1	3	5	8	13	18	30		
	6	1	1	3	5	10	13	22		
	4	1	1	2	4	7	10	16		
	3	1	1	1	3	6	8	14		
	2	1	1	1	3	5	7	11		

XHHH, XHHW, XHHW-2	0	1	1	1	4	5	8
1	0	1	1	1	4	5	8
1/0	0	1	1	1	3	4	7
2/0	0	0	1	1	2	3	6
3/0	0	0	1	1	1	3	5
4/0	0	0	1	1	1	2	4
250	0	0	0	1	1	1	3
300	0	0	0	1	1	1	3
350	0	0	0	1	1	1	2
400	0	0	0	0	1	1	1
500	0	0	0	0	1	1	1
600	0	0	0	0	1	1	1
700	0	0	0	0	1	1	1
750	0	0	0	0	0	1	1
800	0	0	0	0	0	1	1
900	0	0	0	0	0	1	1
1000	0	0	0	0	0	0	1
1250	0	0	0	0	0	0	1
1500	0	0	0	0	0	0	1

Continued

Table C6 Maximum Number of Conductors or Fixture Wires in Liquidtight Flexible Nonmetallic Conduit (Type LFNC-A*) (Based on Table 1) *Continued*

CONDUCTORS

Type	Conductor Size (AWG/kcmil)	Metric Designator (Trade Size)							
		12 (3/8)	16 (1/2)	21 (3/4)	27 (1)	35 (1-1/4)	41 (1-1/2)	53 (2)
XHH, XHHW, XHHW-2	1750	0	0	0	0	0	0	0
	2000	0	0	0	0	0	0	0

Table C7 Maximum Number of Conductors or Fixture Wires in Liquidtight Flexible Metal Conduit (LFMC) (Based on Table 1)

		CONDUCTORS											
	Conductor Size (AWG/kcmil)	Metric Designator (Trade Size)											
Type		16 (1/2)	21 (3/4)	27 (1)	35 (1-1/4)	41 (1-1/2)	53 (2)	63 (2-1/2)	78 (3)	91 (3-1/2)	103 (4)
RHH, RHW, RHW-2	14	4	7	12	21	27	44	66	102	133	173
	12	3	6	10	17	22	36	55	84	110	144
	10	3	5	8	14	18	29	44	68	89	116
	8	1	2	4	7	9	15	23	36	46	61
	6	1	1	3	6	7	12	18	28	37	48
	4	1	1	2	4	6	9	14	22	29	38
	3	1	1	1	4	5	8	13	19	25	33
	2	1	1	1	3	4	7	11	17	22	29
	1	0	1	1	1	3	5	7	11	14	19
	1/0	0	1	1	1	2	4	6	10	13	16
	2/0	0	1	1	1	1	3	5	8	11	14

Continued

Table C7 Maximum Number of Conductors or Fixture Wires in Liquidtight Flexible Metal Conduit (LFMC) (Based on Table 1) *Continued*

	Conductor Size (AWG/kcmil)	Metric Designator (Trade Size)									
Type		16 (1/2)	21 (3/4)	27 (1)	35 (1-1/4)	41 (1-1/2)	53 (2)	63 (2-1/2)	78 (3)	91 (3-1/2)	103 (4)
RHH, RHW, RHW-2	3/0	0	0	1	1	1	3	4	7	9	12
	4/0	0	0	1	1	1	2	4	6	8	10
	250	0	0	0	1	1	1	3	4	6	8
	300	0	0	0	1	1	1	2	4	5	7
	350	0	0	0	1	1	1	2	3	5	6
	400	0	0	0	1	1	1	1	3	4	6
	500	0	0	0	1	1	1	1	3	4	5
	600	0	0	0	0	1	1	1	2	3	4
	700	0	0	0	0	1	0	1	1	3	3
	750	0	0	0	0	0	1	1	1	2	3
	800	0	0	0	0	0	1	1	1	2	3
	900	0	0	0	0	0	1	1	1	2	3

Conductor Type	Size										
(continued)	1000	0	0	0	0	0	1	1	1	1	3
	1250	0	0	0	0	0	0	1	1	1	1
	1500	0	0	0	0	0	0	1	1	1	1
	1750	0	0	0	0	0	0	1	1	1	1
	2000	0	0	0	0	0	0	0	1	1	1
TW, THHW, THW-2	14	9	15	25	44	57	93	140	215	280	365
	12	7	12	19	33	43	71	108	165	215	280
	10	5	9	14	25	32	53	80	123	160	209
	8	3	5	8	14	18	29	44	68	89	116
RHH*, RHW*, RHW-2*	14	6	10	16	29	38	62	93	143	186	243
	12	5	8	13	23	30	50	75	115	149	195
	10	3	6	10	18	23	39	58	89	117	152
	8	1	4	6	11	14	23	35	53	70	91
RHH*, RHW*, RHW-2*, TW, THW, THHW, THW-2	6	1	3	5	8	11	18	27	41	53	70
	4	1	1	3	6	8	13	20	30	40	52
	3	1	1	3	5	7	11	17	26	34	44
	2	1	1	2	4	6	9	14	22	29	38
	1	1	1	1	3	4	7	10	15	20	26

Continued

Table C7 Maximum Number of Conductors or Fixture Wires in Liquidtight Flexible Metal Conduit (LFMC) (Based on Table 1) *Continued*

CONDUCTORS

Type	Conductor Size (AWG/kcmil)	16 (1/2)	21 (3/4)	27 (1)	35 (1-1/4)	41 (1-1/2)	53 (2)	63 (2-1/2)	78 (3)	91 (3-1/2)	103 (4)
RHH*, RHW*, RHW-2*, TW, THW, THHW, THW-2	1/0	0	0	1	2	3	6	8	13	17	23
	2/0	0	1	1	2	3	5	7	11	15	19
	3/0	0	1	1	1	2	4	6	9	12	16
	4/0	0	0	1	1	1	3	5	8	10	13
	250	0	0	0	1	1	3	4	6	8	11
	300	0	0	1	1	1	2	3	5	7	9
	350	0	0	0	1	1	1	3	5	6	8
	400	0	0	0	1	1	1	3	4	6	7
	500	0	0	0	1	1	1	2	3	5	6
	600	0	0	0	1	1	1	1	3	4	5
	700	0	0	0	0	1	1	1	2	3	4
	750	0	0	0	0	1	1	1	2	3	4
	800	0	0	0	0	1	1	1	2	3	4

Metric Designator (Trade Size)

THHN, THWN, THWN-2										
900	0	0	0	0	0	1	1	1	3	3
1000	0	0	0	0	0	1	1	1	2	3
1250	0	0	0	0	0	1	1	1	1	2
1500	0	0	0	0	0	0	1	1	1	2
1750	0	0	0	0	0	0	1	1	1	1
2000	0	0	0	0	0	0	1	1	1	1
14	13	22	36	63	81	133	201	308	401	523
12	9	16	26	46	59	97	146	225	292	381
10	6	10	16	29	37	61	92	141	184	240
8	3	6	9	16	21	35	53	81	106	138
6	2	4	7	12	15	25	38	59	76	100
4	1	2	4	7	9	15	23	36	47	61
3	1	1	3	6	8	13	20	30	40	52
2	1	1	3	5	7	11	17	26	33	44
1	1	1	1	4	5	8	12	19	25	32
1/0	1	1	1	3	4	7	10	16	21	27
2/0	0	1	1	2	3	6	8	13	17	23

Continued

Table C7 Maximum Number of Conductors or Fixture Wires in Liquidtight Flexible Metal Conduit (LFMC) (Based on Table 1) *Continued*

CONDUCTORS

Type	Conductor Size (AWG/kcmil)	Metric Designator (Trade Size)									
		16 (1/2)	21 (3/4)	27 (1)	35 (1-1/4)	41 (1-1/2)	53 (2)	63 (2-1/2)	78 (3)	91 (3-1/2)	103 (4)
THHN, THWN, THWN-2	3/0	0	1	1	1	3	5	7	11	14	19
	4/0	0	1	1	1	2	4	6	9	12	15
	250	0	0	1	1	1	3	5	7	10	12
	300	0	0	1	1	1	3	4	6	8	11
	350	0	0	1	1	1	2	3	5	7	9
	400	0	0	0	1	1	1	3	5	6	8
	500	0	0	0	1	1	1	2	4	5	7
	600	0	0	0	0	1	1	1	3	4	6
	700	0	0	0	1	1	1	1	3	4	5
	750	0	0	0	0	1	1	1	3	3	5
	800	0	0	0	0	1	1	1	2	3	4
	900	0	0	0	0	0	1	1	2	3	4
	1000	0	0	0	0	0	1	1	1	3	3

Type	Size										
FEP, FEPB, PFA, PFAH, TFE	14	12	21	35	61	79	129	195	299	389	507
	12	9	15	25	44	57	94	142	218	284	370
	10	6	11	18	32	41	68	102	156	203	266
	8	3	6	10	18	23	39	58	89	117	152
	6	2	4	7	13	17	27	41	64	83	108
	4	1	3	5	9	12	19	29	44	58	75
	3	1	2	4	7	10	16	24	37	48	63
	2	1	1	3	6	8	13	20	30	40	52
	1	1	1	2	4	5	9	14	21	28	36
PFAH, TFE	1	1	1	2	4	5	9	14	21	28	36
PFA, PFAH, TFE, Z	1/0	1	1	1	3	4	7	11	18	23	30
	2/0	1	1	1	3	4	6	9	14	19	25
	3/0	0	1	1	2	3	5	8	12	16	20
	4/0	0	1	1	1	2	4	6	10	13	17
Z	14	20	26	42	73	95	156	235	360	469	611
	12	14	18	30	52	67	111	167	255	332	434
	10	8	11	18	32	41	68	102	156	203	266

Continued

Table C7 Maximum Number of Conductors or Fixture Wires in Liquidtight Flexible Metal Conduit (LFMC) (Based on Table 1) *Continued*

CONDUCTORS

Type	Conductor Size (AWG/kcmil)	Metric Designator (Trade Size)									
		16 (1/2)	21 (3/4)	27 (1)	35 (1-1/4)	41 (1-1/2)	53 (2)	63 (2-1/2)	78 (3)	91 (3-1/2)	103 (4)
Z	8	5	7	11	20	26	43	64	99	129	168
	6	4	5	8	14	18	30	45	69	90	118
	4	2	3	5	9	12	20	31	48	62	81
	3	2	2	4	7	9	15	23	35	45	59
	2	1	1	3	6	7	12	19	29	38	49
	1	1	1	2	5	6	10	15	23	30	40
XHH, XHHW, XHHW-2, ZW	14	9	15	25	44	57	93	140	215	280	365
	12	7	12	19	33	43	71	108	165	215	280
	10	5	9	14	25	32	53	80	123	160	209
	8	3	5	8	14	18	29	44	68	89	116
	6	1	3	6	10	13	22	33	50	66	86
	4	1	2	4	7	9	16	24	36	48	62

Conductor Size										
XHH, XHHW, XHHW-2										
3	52	40	31	20	13	8	6	3	1	1
2	44	34	26	17	11	7	5	3	1	1
1	33	25	19	12	8	5	4	1	1	1
1/0	28	21	16	10	7	4	3	1	1	1
2/0	23	17	13	9	6	3	2	1	0	0
3/0	19	14	11	7	5	3	1	1	0	0
4/0	16	12	9	6	4	2	1	1	0	0
250	13	10	7	5	3	1	1	1	0	0
300	11	8	6	4	3	1	1	1	0	0
350	10	7	5	3	2	1	1	1	0	0
400	8	6	5	3	1	1	1	0	0	0
500	7	5	4	2	1	1	1	0	0	0
600	6	4	3	1	1	1	1	0	0	0
700	5	4	3	1	1	1	0	0	0	0
750	5	3	3	1	1	1	0	0	0	0
800	4	3	2	1	1	1	0	0	0	0
900	4	3	2	1	1	1	0	0	0	0

Continued

Table C7 Maximum Number of Conductors or Fixture Wires in Liquidtight Flexible Metal Conduit (LFMC) (Based on Table 1) *Continued*

CONDUCTORS

Type	Conductor Size (AWG/kcmil)	Metric Designator (Trade Size)									
		16 (1/2)	21 (3/4)	27 (1)	35 (1-1/4)	41 (1-1/2)	53 (2)	63 (2-1/2)	78 (3)	91 (3-1/2)	103 (4)
XHH, XHHW, XHHW-2	1000	0	0	0	0	0	0	1	1	3	3
	1250	0	0	0	0	0	1	1	1	3	3
	1500	0	0	0	0	0	1	1	1	1	2
	1750	0	0	0	0	0	0	1	1	1	2
	2000	0	0	0	0	0	0	1	1	1	2

*Types RHH, RHW, and RHW-2 without outer covering.

Table C8 Maximum Number of Conductors or Fixture Wires in Rigid Metal Conduit (RMC) (Based on Table 1)

CONDUCTORS

	Conductor Size (AWG/kcmil)	Metric Designator (Trade Size)											
Type		16 (1/2)	21 (3/4)	27 (1)	35 (1-1/4)	41 (1-1/2)	53 (2)	63 (2-1/2)	78 (3)	91 (3-1/2)	103 (4)	129 (5)	155 (6)
RHH, RHW, RHW-2	14	4	7	12	21	28	46	66	102	136	176	276	398
	12	3	6	10	17	23	38	55	85	113	146	229	330
	10	3	5	8	14	19	31	44	68	91	118	185	267
	8	1	2	4	7	10	16	23	36	48	61	97	139
	6	1	1	3	6	8	13	18	29	38	49	77	112
	4	1	1	2	4	6	10	14	22	30	38	60	87
	3	1	1	2	4	5	9	12	19	26	34	53	76
	2	1	1	1	3	4	7	11	17	23	29	46	66
	1	0	1	1	1	3	5	7	11	15	19	30	44
	1/0	0	1	1	1	2	4	6	10	13	17	26	38
	2/0	0	1	1	1	2	4	5	8	11	14	23	33

Continued

Table C8 Maximum Number of Conductors or Fixture Wires in Rigid Metal Conduit (RMC) (Based on Table 1) *Continued*

CONDUCTORS

Type	Conductor Size (AWG/kcmil)	Metric Designator (Trade Size)											
		16 (1/2)	21 (3/4)	27 (1)	35 (1-1/4)	41 (1-1/2)	53 (2)	63 (2-1/2)	78 (3)	91 (3-1/2)	103 (4)	129 (5)	155 (6)
RHH, RHW, RHW-2	3/0	0	0	1	1	1	3	4	7	10	12	20	28
	4/0	0	0	1	1	1	3	4	6	8	11	17	24
	250	0	0	0	1	1	1	3	4	6	8	13	18
	300	0	0	0	1	1	1	2	4	5	7	11	16
	350	0	0	0	1	1	1	2	4	5	6	10	15
	400	0	0	0	1	1	1	1	3	4	6	9	13
	500	0	0	0	1	1	1	1	3	4	5	8	11
	600	0	0	0	0	1	1	1	2	3	4	6	9
	700	0	0	0	0	1	1	1	1	3	4	6	8
	750	0	0	0	0	0	1	1	1	3	3	5	8
	800	0	0	0	0	0	1	1	1	2	3	5	7
	900	0	0	0	0	0	1	1	1	2	3	5	7

Type	Size												
	1000	0	0	0	0	0	1	1	1	1	3	4	6
	1250	0	0	0	0	0	0	1	1	1	1	3	5
	1500	0	0	0	0	0	0	1	1	1	1	3	4
	1750	0	0	0	0	0	0	1	1	1	1	2	4
	2000	0	0	0	0	0	0	0	1	1	1	2	3
TW, THW, THW-2	14	9	15	25	44	59	98	140	216	288	370	581	839
	12	7	12	19	33	45	75	107	165	221	284	446	644
	10	5	9	14	25	34	56	80	123	164	212	332	480
	8	3	5	8	14	19	31	44	68	91	118	185	267
RHH*, RHW*, RHW-2*	14	6	10	17	29	39	65	93	143	191	246	387	558
RHH*, RHW*, RHW-2*	12	5	8	13	23	32	52	75	115	154	198	311	448
	10	3	6	10	18	25	41	58	90	120	154	242	350
RHH*, RHW*, RHW-2*	8	1	4	6	11	15	24	35	54	72	92	145	209

Continued

Table C8 Maximum Number of Conductors or Fixture Wires in Rigid Metal Conduit (RMC) (Based on Table 1) *Continued*

CONDUCTORS

Type	Conductor Size (AWG/kcmil)	Metric Designator (Trade Size)											
		16 (1/2)	21 (3/4)	27 (1)	35 (1-1/4)	41 (1-1/2)	53 (2)	63 (2-1/2)	78 (3)	91 (3-1/2)	103 (4)	129 (5)	155 (6)
RHH*, RHW*, RHW-2*	6	1	3	5	8	11	18	27	41	55	71	111	160
TW, THW, THHW, THW-2	4	1	1	3	6	8	14	20	31	41	53	83	120
	3	1	1	3	5	7	12	17	26	35	45	71	103
	2	1	1	2	4	6	10	14	22	30	38	60	87
	1	1	1	1	3	4	7	10	15	21	27	42	61
	1/0	0	1	1	2	3	6	8	13	18	23	36	52
	2/0	0	1	1	2	3	5	7	11	15	19	31	44
	3/0	0	1	1	1	2	4	6	9	13	16	26	37
	4/0	0	0	1	1	1	3	5	8	10	14	21	31
	250	0	0	1	1	1	3	4	6	8	11	17	25
	300	0	0	1	1	1	2	3	5	7	9	15	22
	350	0	0	0	1	1	1	3	5	6	8	13	19

Size												
400	17	12	7	6	4	3	1	1	1	0	0	0
500	14	10	6	5	3	2	1	1	1	0	0	0
600	12	8	5	4	3	1	1	1	1	0	0	0
700	10	7	4	3	2	1	1	0	0	0	0	0
750	10	7	4	3	2	1	1	0	0	0	0	0
800	9	6	4	3	2	1	1	0	0	0	0	0
900	8	6	4	3	1	1	1	0	0	0	0	0
1000	8	5	3	2	1	1	1	0	0	0	0	0
1250	6	4	2	1	1	1	1	0	0	0	0	0
1500	5	3	2	1	1	1	1	0	0	0	0	0
1750	4	3	1	1	1	1	0	0	0	0	0	0
2000	4	3	1	1	1	1	0	0	0	0	0	0
14	1202	833	531	412	309	200	140	85	63	36	22	13
12	877	608	387	301	225	146	102	62	46	26	16	9
10	552	383	244	189	142	92	64	39	29	17	10	6
8	318	221	140	109	82	53	37	22	16	9	6	3
6	230	159	101	79	59	38	27	16	12	7	4	2

THHN, THWN, THWN-2

Continued

Table C8 Maximum Number of Conductors or Fixture Wires in Rigid Metal Conduit (RMC) (Based on Table 1) *Continued*

CONDUCTORS

Type	Conductor Size (AWG/kcmil)	Metric Designator (Trade Size)											
		16 (1/2)	21 (3/4)	27 (1)	35 (1-1/4)	41 (1-1/2)	53 (2)	63 (2-1/2)	78 (3)	91 (3-1/2)	103 (4)	129 (5)	155 (6)
THHN, THWN, THWN-2	4	1	2	4	7	10	16	23	36	48	62	98	141
	3	1	1	3	6	8	14	20	31	41	53	83	120
	2	1	1	3	5	7	11	17	26	34	44	70	100
	1	1	1	1	4	5	8	12	19	25	33	51	74
	1/0	1	1	1	3	4	7	10	16	21	27	43	63
	2/0	0	1	1	2	3	6	8	13	18	23	36	52
	3/0	0	1	1	1	3	5	7	11	15	19	30	43
	4/0	0	1	1	1	2	4	6	9	12	16	25	36
	250	0	0	1	1	1	3	5	7	10	13	20	29
	300	0	0	1	1	1	3	4	6	8	11	17	25
	350	0	0	1	1	1	2	3	5	7	10	15	22
	400	0	0	1	1	1	2	3	5	7	8	13	20
	500	0	0	0	1	1	1	2	4	5	7	11	16

Type	Size												
	600	13	9	6	4	3	1	1	1	1	0	0	0
	700	11	8	5	4	3	1	1	1	1	0	0	0
	750	11	7	5	4	3	1	1	1	0	0	0	0
	800	10	7	4	3	2	1	1	1	0	0	0	0
	900	9	6	4	3	2	1	1	1	0	0	0	0
	1000	8	6	4	3	1	1	1	1	0	0	0	0
FEP, FEPB, PFA, PFAH, TFE	14	1166	808	515	400	300	194	136	83	61	35	22	12
	12	851	590	376	292	219	142	99	60	44	26	16	9
	10	610	423	269	209	157	102	71	43	32	18	11	6
	8	350	242	154	120	90	58	41	25	18	10	6	3
	6	249	172	110	85	64	41	29	17	13	7	4	2
	4	174	120	77	59	44	29	20	12	9	5	3	1
	3	145	100	64	50	37	24	17	10	7	4	2	1
	2	120	83	53	41	31	20	14	8	6	3	1	1
PFA, PFAH, TFE	1	83	57	37	28	21	14	9	6	4	2	1	1
PFA, PFAH, TFE, Z	1/0	69	48	30	24	18	11	8	5	3	1	1	1
	2/0	57	40	25	19	14	9	6	4	3	1	1	1

Continued

Table C8 Maximum Number of Conductors or Fixture Wires in Rigid Metal Conduit (RMC) (Based on Table 1) *Continued*

CONDUCTORS

Type	Conductor Size (AWG/kcmil)	Metric Designator (Trade Size)											
		16 (1/2)	21 (3/4)	27 (1)	35 (1-1/4)	41 (1-1/2)	53 (2)	63 (2-1/2)	78 (3)	91 (3-1/2)	103 (4)	129 (5)	155 (6)
PFA, PFAH, TFE, Z	3/0	0	1	1	2	3	5	8	12	16	21	33	47
	4/0	0	1	1	1	2	4	6	10	13	17	27	39
Z	14	15	26	42	73	100	164	234	361	482	621	974	1405
	12	10	18	30	52	71	116	166	256	342	440	691	997
	10	6	11	18	32	43	71	102	157	209	269	423	610
	8	4	7	11	20	27	45	64	99	132	170	267	386
	6	3	5	8	14	19	31	45	69	93	120	188	271
	4	1	3	5	9	13	22	31	48	64	82	129	186
	3	1	2	4	7	9	16	22	35	47	60	94	136
	2	1	1	3	6	8	13	19	29	39	50	78	113
	1	1	1	2	5	6	10	15	23	31	40	63	92

Type	Size												
XHH, XHHW, XHHW-2, ZW	14	9	15	25	44	59	98	140	216	288	370	581	839
	12	7	12	19	33	45	75	107	165	221	284	446	644
	10	5	9	14	25	34	56	80	123	164	212	332	480
	8	3	5	8	14	19	31	44	68	91	118	185	267
	6	1	3	6	10	14	23	33	51	68	87	137	197
	4	1	2	4	7	10	16	24	37	49	63	99	143
	3	1	1	3	6	8	14	20	31	41	53	84	121
	2	1	1	3	5	7	12	17	26	35	45	70	101
	1	1	1	1	4	5	9	12	19	26	33	52	76
XHH, XHHW, XHHW-2	1/0	1	1	1	3	4	7	10	16	22	28	44	64
	2/0	0	1	1	2	3	6	9	13	18	23	37	53
	3/0	0	1	1	1	3	5	7	11	15	19	30	44
	4/0	0	1	1	1	2	4	6	9	12	16	25	36
	250	0	0	1	1	1	3	5	7	10	13	20	30
	300	0	0	1	1	1	3	4	6	9	11	18	25
	350	0	0	1	1	1	2	3	6	7	10	15	22
	400	0	0	1	1	1	2	3	5	7	9	14	20
	500	0	0	0	1	1	1	2	4	5	7	11	16

Continued
Continued

Table C8 Maximum Number of Conductors or Fixture Wires in Rigid Metal Conduit (RMC) (Based on Table 1)

CONDUCTORS

Type	Conductor Size (AWG/kcmil)	Metric Designator (Trade Size)											
		16 (1/2)	21 (3/4)	27 (1)	35 (1-1/4)	41 (1-1/2)	53 (2)	63 (2-1/2)	78 (3)	91 (3-1/2)	103 (4)	129 (5)	155 (6)
XHH, XHHW, XHHW-2	600	0	0	0	1	1	1	1	3	4	6	9	13
	700	0	0	0	1	1	1	1	3	4	5	8	11
	750	0	0	0	0	1	1	1	3	4	5	7	11
	800	0	0	0	0	1	1	1	2	3	4	7	10
	900	0	0	0	0	1	1	1	2	3	4	6	9
	1000	0	0	0	0	1	1	1	1	3	4	6	8
	1250	0	0	0	0	0	1	1	1	2	3	4	6
	1500	0	0	0	0	0	1	1	1	1	2	4	5
	1750	0	0	0	0	0	0	1	1	1	1	3	5
	2000	0	0	0	0	0	0	1	1	1	1	3	4

Table C9 Maximum Number of Conductors or Fixture Wires in Rigid PVC Conduit, Schedule 80 (Based on Table 1)

Type	Conductor Size (AWG/kcmil)	CONDUCTORS Metric Designator (Trade Size)												
		16 (1/2)	21 (3/4)	27 (1)	35 (1-1/4)	41 (1-1/2)	53 (2)	63 (2-1/2)	78 (3)	91 (3-1/2)	103 (4)	129 (5)	155 (6)
RHH, RHW, RHW-2	14	3	5	9	17	23	39	56	88	118	153	243	349
	12	2	4	7	14	19	32	46	73	98	127	202	290
	10	1	3	6	11	15	26	37	59	79	103	163	234
	8	1	1	3	6	8	13	19	31	41	54	85	122
	6	1	1	2	4	6	11	16	24	33	43	68	98
	4	1	1	1	3	5	8	12	19	26	33	53	77
	3	0	1	1	3	4	7	11	17	23	29	47	67
	2	0	1	1	3	4	6	9	14	20	25	41	58
	1	0	0	1	1	2	4	6	9	13	17	27	38
	1/0	0	0	1	1	1	3	5	8	11	15	23	33
	2/0	0	0	1	1	1	3	4	7	10	13	20	29

Continued

Table C9 Maximum Number of Conductors or Fixture Wires in Rigid PVC Conduit, Schedule 80 (Based on Table 1) *Continued*

CONDUCTORS

Type	Conductor Size (AWG/kcmil)	Metric Designator (Trade Size)											
		16 (1/2)	21 (3/4)	27 (1)	35 (1-1/4)	41 (1-1/2)	53 (2)	63 (2-1/2)	78 (3)	91 (3-1/2)	103 (4)	129 (5)	155 (6)
RHH, RHW, RHW-2	3/0	0	0	1	1	1	3	4	6	8	11	17	25
	4/0	0	0	0	1	1	2	3	5	7	9	15	21
	250	0	0	0	1	1	1	2	4	5	7	11	16
	300	0	0	0	1	1	1	2	3	5	6	10	14
	350	0	0	0	1	1	1	1	3	4	5	9	13
	400	0	0	0	0	1	1	1	3	4	5	8	12
	500	0	0	0	0	1	1	1	2	3	4	7	10
	600	0	0	0	0	0	1	1	1	3	3	6	8
	700	0	0	0	0	0	1	1	1	2	3	5	7
	750	0	0	0	0	0	1	1	1	2	3	5	7
	800	0	0	0	0	0	1	1	1	2	3	4	7

Type	Size												
	1000	0	0	0	0	0	1	1	1	1	2	4	5
	1250	0	0	0	0	0	0	1	1	1	1	3	4
	1500	0	0	0	0	0	0	1	1	1	1	2	4
	1750	0	0	0	0	0	0	0	1	1	1	2	3
	2000	0	0	0	0	0	0	0	1	1	1	1	3
TW, THHW, THW, THW-2	14	6	11	20	35	49	82	118	185	250	324	514	736
	12	5	9	15	27	38	63	91	142	192	248	394	565
	10	3	6	11	20	28	47	67	106	143	185	294	421
	8	1	3	6	11	15	26	37	59	79	103	163	234
RHH*, RHW*, RHW-2*	14	4	8	13	23	32	55	79	123	166	215	341	490
	12	3	6	10	19	26	44	63	99	133	173	274	394
	10	2	5	8	15	20	34	49	77	104	135	214	307
	8	1	3	5	9	12	20	29	46	62	81	128	184
RHH*, RHW*, RHW-2*, TW, THW, THHW, THW-2	6	1	1	3	7	9	16	22	35	48	62	98	141
	4	1	1	3	5	7	12	17	26	35	46	73	105
	3	1	1	2	4	6	10	14	22	30	39	63	90
	2	1	1	1	3	5	8	12	19	26	33	53	77
	1	0	1	1	2	3	6	8	13	18	23	37	54

Continued

Table C9 Maximum Number of Conductors or Fixture Wires in Rigid PVC Conduit, Schedule 80 (Based on Table 1) *Continued*

CONDUCTORS

Type	Conductor Size (AWG/kcmil)	Metric Designator (Trade Size)											
		16 (1/2)	21 (3/4)	27 (1)	35 (1-1/4)	41 (1-1/2)	53 (2)	63 (2-1/2)	78 (3)	91 (3-1/2)	103 (4)	129 (5)	155 (6)
RHH*, RHW*, RHW-2*, TW, THW, THHW, THW-2	1/0	0	1	1	1	2	5	7	11	15	20	32	46
	2/0	0	1	1	1	2	4	6	10	13	17	27	39
	3/0	0	0	1	1	1	3	5	8	11	14	23	33
	4/0	0	0	1	1	1	3	4	7	9	12	19	27
	250	0	0	0	1	1	2	3	5	7	9	15	22
	300	0	0	0	1	1	1	3	5	6	8	13	19
	350	0	0	0	1	1	1	2	4	6	7	12	17
	400	0	0	0	1	1	1	2	4	5	7	10	15
	500	0	0	0	1	1	1	1	3	4	5	9	13
	600	0	0	0	0	1	1	1	2	3	4	7	10
	700	0	0	0	0	1	1	1	2	3	4	6	9
	750	0	0	0	0	0	1	1	1	3	4	6	8

Size												
800	0	0	0	0	0	1	1	1	3	3	6	8
900	0	0	0	0	0	1	1	1	2	3	5	7
1000	0	0	0	0	1	1	1	1	2	3	5	7
1250	0	0	0	0	1	1	1	1	1	2	4	5
1500	0	0	0	0	1	1	1	1	1	1	3	4
1750	0	0	0	0	0	1	1	1	1	1	3	4
2000	0	0	0	0	0	0	1	1	1	1	2	3
THHN, THWN, THWN-2												
14	9	17	28	51	70	118	170	265	358	464	736	1055
12	6	12	20	37	51	86	124	193	261	338	537	770
10	4	7	13	23	32	54	78	122	164	213	338	485
8	2	4	7	13	18	31	45	70	95	123	195	279
6	1	3	5	9	13	22	32	51	68	89	141	202
4	1	1	3	6	8	14	20	31	42	54	86	124
3	1	1	3	5	7	12	17	26	35	46	73	105
2	1	1	2	4	6	10	14	22	30	39	61	88
1	0	1	1	3	4	7	10	16	22	29	45	65
1/0	0	1	1	2	3	6	9	14	18	24	38	55
2/0	0	1	1	1	3	5	7	11	15	20	32	46

Continued

Table C9 Maximum Number of Conductors or Fixture Wires in Rigid PVC Conduit, Schedule 80 (Based on Table 1) *Continued*

CONDUCTORS

Type	Conductor Size (AWG/kcmil)	Metric Designator (Trade Size)											
		16 (1/2)	21 (3/4)	27 (1)	35 (1-1/4)	41 (1-1/2)	53 (2)	63 (2-1/2)	78 (3)	91 (3-1/2)	103 (4)	129 (5)	155 (6)
THHN, THWN, THWN-2	3/0	0	1	1	1	2	4	6	9	13	17	26	38
	4/0	0	0	1	1	1	3	5	8	10	14	22	31
	250	0	0	1	1	1	3	4	6	8	11	18	25
	300	0	0	0	1	1	2	3	5	7	9	15	22
	350	0	0	0	1	1	1	3	5	6	8	13	19
	400	0	0	0	1	1	1	3	4	6	7	12	17
	500	0	0	0	1	1	1	2	3	5	6	10	14
	600	0	0	0	0	1	1	1	3	4	5	8	12
	700	0	0	0	0	1	1	1	2	3	4	7	10
	750	0	0	0	0	1	1	1	2	3	4	7	9
	800	0	0	0	0	0	1	1	2	3	4	6	9
	900	0	0	0	0	0	1	1	1	2	3	6	8
	1000	0	0	0	0	0	1	1	1	2	3	5	7

Type	Size												
FEP, FEPB, PFA, PFAH, TFE	14	8	16	27	49	68	115	164	257	347	450	714	1024
	12	6	12	20	36	50	84	120	188	253	328	521	747
	10	4	8	14	26	36	60	86	135	182	235	374	536
	8	2	5	8	15	20	34	49	77	104	135	214	307
	6	1	3	6	10	14	24	35	55	74	96	152	218
PFA, PFAH, TFE	4	1	2	4	7	10	17	24	38	52	67	106	153
	3	1	1	3	6	8	14	20	32	43	56	89	127
	2	1	1	3	5	7	12	17	26	35	46	73	105
PFA, PFAH, TFE	1	1	1	1	3	5	8	11	18	25	32	51	73
PFA, PFAH, TFE, Z	1/0	0	1	1	3	4	7	10	15	20	27	42	61
	2/0	0	1	1	2	3	5	8	12	17	22	35	50
	3/0	0	1	1	1	2	4	6	10	14	18	29	41
	4/0	0	0	1	1	1	4	5	8	11	15	24	34
Z	14	10	19	33	59	82	138	198	310	418	542	860	1233
	12	7	14	23	42	58	98	141	220	297	385	610	875
	10	4	8	14	26	36	60	86	135	182	235	374	536

Continued

Table C9 Maximum Number of Conductors or Fixture Wires
in Rigid PVC Conduit, Schedule 80 (Based on Table 1) *Continued*

CONDUCTORS

Type	Conductor Size (AWG/kcmil)	Metric Designator (Trade Size)											
		16 (1/2)	21 (3/4)	27 (1)	35 (1-1/4)	41 (1-1/2)	53 (2)	63 (2-1/2)	78 (3)	91 (3-1/2)	103 (4)	129 (5)	155 (6)
Z	8	3	5	9	16	22	38	54	85	115	149	236	339
	6	2	4	6	11	16	26	38	60	81	104	166	238
	4	1	2	4	8	11	18	26	41	55	72	114	164
	3	1	2	3	5	8	13	19	30	40	52	83	119
	2	1	1	2	5	6	11	16	25	33	43	69	99
	1	0	1	2	4	5	9	13	20	27	35	56	80
XHH, XHHW, XHHW-2, ZW	14	6	11	20	35	49	82	118	185	250	324	514	736
	12	5	9	15	27	38	63	91	142	192	248	394	565
	10	3	6	11	20	28	47	67	106	143	185	294	421
	8	1	3	6	11	15	26	37	59	79	103	163	234
	6	1	2	4	8	11	19	28	43	59	76	121	173

4	125	87	55	42	31	20	14	8	6	3	1	1
3	106	74	47	36	26	17	12	7	5	3	1	1
2	89	62	39	30	22	14	10	6	4	2	1	1
1	66	46	29	22	16	10	7	4	3	1	1	0
1/0	56	39	24	19	14	9	6	3	2	1	1	0
2/0	46	32	20	16	11	7	5	3	1	1	1	0
3/0	38	27	17	13	9	6	4	2	1	1	1	0
4/0	32	22	14	11	8	5	3	1	1	1	0	0
250	26	18	11	9	6	4	3	1	1	1	0	0
300	22	15	10	7	5	3	2	1	1	1	0	0
350	20	14	8	6	5	3	1	1	1	0	0	0
400	17	12	7	6	4	3	1	1	1	0	0	0
500	14	10	6	5	3	2	1	1	1	0	0	0
600	11	8	5	4	3	1	1	1	0	0	0	0
700	10	7	4	3	2	1	1	1	0	0	0	0
750	9	6	4	3	2	1	1	1	0	0	0	0

XHH, XHHW, XHHW-2

Continued

Table C9 Maximum Number of Conductors or Fixture Wires in Rigid PVC Conduit, Schedule 80 (Based on Table 1) *Continued*

CONDUCTORS

Type	Conductor Size (AWG/kcmil)	Metric Designator (Trade Size)											
		16 (1/2)	21 (3/4)	27 (1)	35 (1-1/4)	41 (1-1/2)	53 (2)	63 (2-1/2)	78 (3)	91 (3-1/2)	103 (4)	129 (5)	155 (6)
XHH, XHHW, XHHW-2	800	0	0	0	0	1	1	1	1	3	4	6	9
	900	0	0	0	0	1	1	1	—	3	3	5	8
	1000	0	0	0	0	0	1	1	1	2	3	5	7
	1250	0	0	0	0	0	1	1	1	1	2	4	6
	1500	0	0	0	0	0	0	1	1	1	1	3	5
	1750	0	0	0	0	0	0	1	1	1	1	3	4
	2000	0	0	0	0	0	0	1	1	1	1	2	4

Table C10 Maximum Number of Conductors or Fixture Wires in Rigid PVC Conduit, Schedule 40 and HDPE Conduit (Based on Table 1)

		CONDUCTORS												
						Metric Designator (Trade Size)								
Type	Conductor Size (AWG/kcmil)	16 (1/2)	21 (3/4)	27 (1)	35 (1-1/4)	41 (1-1/2)	53 (2)	63 (2-1/2)	78 (3)	91 (3-1/2)	103 (4)	129 (5)	155 (6)
RHH, RHW, RHW-2	14	4	7	11	20	27	45	64	99	133	171	269	390
	12	3	5	9	16	22	37	53	82	110	142	224	323
	10	2	4	7	13	18	30	43	66	89	115	181	261
	8	1	2	4	7	9	15	22	35	46	60	94	137
	6	1	1	3	5	7	12	18	28	37	48	76	109
	4	1	1	2	4	6	10	14	22	29	37	59	85
	3	1	1	1	4	5	8	12	19	25	33	52	75
	2	1	1	1	3	4	7	10	16	22	28	45	65
	1	0	1	1	1	3	5	7	11	14	19	29	43
	1/0	0	1	1	1	2	4	6	9	13	16	26	37
	2/0	0	0	1	1	2	3	5	8	11	14	22	32

Continued

Table C10 Maximum Number of Conductors or Fixture Wires in Rigid PVC Conduit, Schedule 40 and HDPE Conduit (Based on Table 1) *Continued*

CONDUCTORS

Type	Conductor Size (AWG/kcmil)	Metric Designator (Trade Size)											
		16 (1/2)	21 (3/4)	27 (1)	35 (1-1/4)	41 (1-1/2)	53 (2)	63 (2-1/2)	78 (3)	91 (3-1/2)	103 (4)	129 (5)	155 (6)
RHH, RHW, RHW-2	3/0	0	0	1	1	1	3	4	7	9	12	19	28
	4/0	0	0	1	1	1	2	4	6	8	10	16	24
	250	0	0	0	1	1	1	3	4	6	8	12	18
	300	0	0	0	1	1	1	2	4	5	7	11	16
	350	0	0	0	1	1	1	2	3	5	6	10	14
	400	0	0	0	1	1	1	1	3	4	6	9	13
	500	0	0	0	0	1	1	1	3	4	5	8	11
	600	0	0	0	0	1	1	1	2	3	4	6	9
	700	0	0	0	0	0	1	1	1	3	3	6	8
	750	0	0	0	0	0	1	1	1	2	3	5	8
	800	0	0	0	0	0	1	1	1	2	3	5	7
	900	0	0	0	0	0	1	1	1	2	3	5	7

Type	Size												
	1000	0	0	0	0	0	1	1	1	1	3	4	6
	1250	0	0	0	0	0	0	1	1	1	1	3	5
	1500	0	0	0	0	0	0	1	1	1	1	3	4
	1750	0	0	0	0	0	0	1	1	1	1	2	3
	2000	0	0	0	0	0	0	0	1	1	1	2	3
TW, THHW, THW, THW-2	14	8	14	24	42	57	94	135	209	280	361	568	822
	12	6	11	18	32	44	72	103	160	215	277	436	631
	10	4	8	13	24	32	54	77	119	160	206	325	470
	8	2	4	7	13	18	30	43	66	89	115	181	261
RHH*, RHW*, RHW-2*	14	5	9	16	28	38	63	90	139	186	240	378	546
	12	4	8	12	22	30	50	72	112	150	193	304	439
	10	3	6	10	17	24	39	56	87	117	150	237	343
	8	1	3	6	10	14	23	33	52	70	90	142	205
TW, THW, THHW, THW-2	6	1	2	4	8	11	18	26	40	53	69	109	157
	4	1	1	3	6	8	13	19	30	40	51	81	117
	3	1	1	3	5	7	11	16	25	34	44	69	100
	2	1	1	2	4	6	10	14	22	29	37	59	85
	1	0	1	1	3	4	7	10	15	20	26	41	60

Continued

Table C10 Maximum Number of Conductors or Fixture Wires in Rigid PVC Conduit, Schedule 40 and HDPE Conduit (Based on Table 1) *Continued*

CONDUCTORS

Type	Conductor Size (AWG/kcmil)	Metric Designator (Trade Size)											
		16 (1/2)	21 (3/4)	27 (1)	35 (1-1/4)	41 (1-1/2)	53 (2)	63 (2-1/2)	78 (3)	91 (3-1/2)	103 (4)	129 (5)	155 (6)
TW, THW, THHW, THW-2	1/0	0	1	1	2	3	6	8	13	17	22	35	51
	2/0	0	1	1	1	3	5	7	11	15	19	30	43
	3/0	0	1	1	1	2	4	6	9	12	16	25	36
	4/0	0	0	1	1	1	3	5	8	10	13	21	30
	250	0	0	1	1	1	3	4	6	8	11	17	25
	300	0	0	1	1	1	2	3	5	7	9	15	21
	350	0	0	0	1	1	1	3	5	6	8	13	19
	400	0	0	0	1	1	1	3	4	6	7	12	17
	500	0	0	0	1	1	1	2	3	5	6	10	14
	600	0	0	0	0	1	1	1	3	4	5	8	11
	700	0	0	0	0	1	1	1	2	3	4	7	10
	750	0	0	0	0	1	1	1	2	3	4	6	10

THHN, THWN, THWN-2												
800	9	6	4	3	2	1	1	1	0	0	0	0
900	8	6	3	3	1	1	1	0	0	0	0	0
1000	7	5	3	2	1	1	1	0	0	0	0	0
1250	6	4	2	1	1	1	1	0	0	0	0	0
1500	5	3	1	1	1	1	1	0	0	0	0	0
1750	4	3	1	1	1	1	0	0	0	0	0	0
2000	4	3	1	1	1	1	0	0	0	0	0	0
14	1178	815	517	401	299	193	135	82	60	34	21	11
12	859	594	377	293	218	141	99	59	43	25	15	8
10	541	374	238	184	137	89	62	37	27	15	9	5
8	312	216	137	106	79	51	36	21	16	9	5	3
6	225	156	99	77	57	37	26	15	11	6	4	1
4	138	96	61	47	35	22	16	9	7	4	2	1
3	117	81	51	40	30	19	13	8	6	3	1	1
2	98	68	43	33	25	16	11	7	5	3	1	1
1	73	50	32	25	18	12	8	5	3	1	1	1
1/0	61	42	27	21	15	10	7	4	3	1	1	1
2/0	51	35	22	17	13	8	6	3	2	1	1	0

Continued

Table C10 Maximum Number of Conductors or Fixture Wires in Rigid PVC Conduit, Schedule 40 and HDPE Conduit (Based on Table 1) *Continued*

CONDUCTORS

Type	Conductor Size (AWG/kcmil)	16 (1/2)	21 (3/4)	27 (1)	35 (1-1/4)	41 (1-1/2)	53 (2)	63 (2-1/2)	78 (3)	91 (3-1/2)	103 (4)	129 (5)	155 (6)
						Metric Designator (Trade Size)							
THHN, THWN, THWN-2	3/0	0	1	1	1	3	5	7	11	14	18	29	42
	4/0	0	1	1	1	2	4	6	9	12	15	24	35
	250	0	0	1	1	1	3	4	7	10	12	20	28
	300	0	0	1	1	1	3	4	6	8	11	17	24
	350	0	0	1	1	1	2	3	5	7	9	15	21
	400	0	0	0	1	1	1	3	5	6	8	13	19
	500	0	0	0	1	1	1	2	4	5	7	11	16
	600	0	0	0	0	1	1	1	3	4	5	9	13
	700	0	0	0	0	1	1	1	3	4	5	8	11
	750	0	0	0	0	1	1	1	2	3	4	7	11
	800	0	0	0	0	1	1	1	2	3	4	7	10
	900	0	0	0	0	1	1	1	2	3	4	6	9
	1000	0	0	0	0	0	1	1	1	3	3	6	8

FEP, FEPB, PFA, PFAH, TFE	14	11	20	33	58	79	131	188	290	389	502	790	1142	
	12	8	15	24	42	58	96	137	212	284	366	577	834	
	10	6	10	17	30	41	69	98	152	204	263	414	598	
	8	3	6	10	17	24	39	56	87	117	150	237	343	
PFA, PFAH, TFE	6	2	4	7	12	17	28	40	62	83	107	169	244	
	4	1	3	5	8	12	19	28	43	58	75	118	170	
	3	1	2	4	7	10	16	23	36	48	62	98	142	
	2	1	1	3	6	8	13	19	30	40	51	81	117	
PFA, PFAH, TFE	1	1	1	2	4	5	9	13	20	28	36	56	81	
PFA, PFAH, TFE, Z	1/0	1	1	1	3	4	8	11	17	23	30	47	68	
	2/0	0	1	1	3	4	6	9	14	19	24	39	56	
	3/0	0	1	1	2	3	5	7	12	16	20	32	46	
	4/0	0	1	1	1	2	4	6	9	13	16	26	38	
Z	14	13	24	40	70	95	158	226	350	469	605	952	1376	
	12	9	17	28	49	68	112	160	248	333	429	675	976	
	10	6	10	17	30	41	69	98	152	204	263	414	598	
	8	3	6	11	19	26	43	62	96	129	166	261	378	

Continued

Table C10　Maximum Number of Conductors or Fixture Wires
in Rigid PVC Conduit, Schedule 40 and HDPE Conduit (Based on Table 1)　*Continued*

CONDUCTORS

Type	Conductor Size (AWG/kcmil)	Metric Designator (Trade Size)											
		16 (1/2)	21 (3/4)	27 (1)	35 (1-1/4)	41 (1-1/2)	53 (2)	63 (2-1/2)	78 (3)	91 (3-1/2)	103 (4)	129 (5)	155 (6)
Z	6	2	4	7	13	18	30	43	67	90	116	184	265
	4	1	3	5	9	12	21	30	46	62	80	126	183
	3	1	2	4	6	9	15	22	34	45	58	92	133
	2	1	1	3	5	7	12	18	28	38	49	77	111
	1	1	1	2	4	6	10	14	23	30	39	62	90
XHH, XHHW, XHHW-2, ZW	14	8	14	24	42	57	94	135	209	280	361	568	822
	12	6	11	18	32	44	72	103	160	215	277	436	631
	10	4	8	13	24	32	54	77	119	160	206	325	470
	8	2	4	7	13	18	30	43	66	89	115	181	261
	6	1	3	5	10	13	22	32	49	66	85	134	193
	4	1	2	4	7	9	16	23	35	48	61	97	140
	3	1	1	3	6	8	13	19	30	40	52	82	118
	2	1	1	3	5	7	11	16	25	34	44	69	99

XHH, XHHW, XHHW-2	74	51	32	25	19	12	8	5	3	1	1	1	1
1/0	62	43	27	21	16	10	7	4	3	1	1	1	1
2/0	52	36	23	17	13	8	6	3	2	1	1	0	0
3/0	43	30	19	14	11	7	5	3	1	1	1	0	0
4/0	35	24	15	12	9	6	4	2	1	1	1	1	0
250	29	20	13	10	7	5	3	1	1	1	0	0	0
300	25	17	11	8	6	4	3	1	1	1	0	0	0
350	22	15	9	7	6	3	2	1	1	0	0	0	0
400	19	13	8	6	5	3	1	1	1	0	0	0	0
500	16	11	7	5	4	2	1	1	1	0	0	0	0
600	13	9	5	4	3	1	1	1	1	0	0	0	0
700	11	8	5	4	3	1	1	1	0	0	0	0	0
750	11	7	4	3	2	1	1	1	0	0	0	0	0
800	10	7	4	3	2	1	1	1	0	0	0	0	0
900	9	6	4	3	2	1	1	1	0	0	0	0	0
1000	8	6	3	3	1	1	1	0	0	0	0	0	0
1250	6	4	3	1	1	1	1	0	0	0	0	0	0

Continued

Table C10 Maximum Number of Conductors or Fixture Wires
in Rigid PVC Conduit, Schedule 40 and HDPE Conduit (Based on Table 1) *Continued*

CONDUCTORS

Type	Conductor Size (AWG/kcmil)	Metric Designator (Trade Size)										
		16 (1/2)	21 (3/4)	27 (1)	35 (1-1/4)	41 (1-1/2)	53 (2)	63 (2-1/2)	78 (3)	91 (3-1/2)	103 (4)	129 (5)	155 (6)
XHH, XHHW, XHHW-2	1500	0	0	0	0	0	1	1	1	1	2	4	5
	1750	0	0	0	0	0	0	1	1	1	1	3	5
	2000	0	0	0	0	0	0	1	1	1	1	3	4

INDEX